Climate Change and Security

**Recent Title in
Security and the Environment**

Arctic Doom, Arctic Boom: The Geopolitics of Climate Change
in the Arctic
Barry Scott Zellen

Climate Change and Security

A Gathering Storm of Global Challenges

Christian Webersik

Security and the Environment
P. H. Liotta, Series Editor

 PRAEGER

AN IMPRINT OF ABC-CLIO, LLC
Santa Barbara, California • Denver, Colorado • Oxford, England

Copyright 2010 by Christian Webersik

Library of Congress Cataloging-in-Publication Data
Webersik, Christian, 1970-
   Climate change and security : a gathering storm of global challenges / Christian Webersik.
      p. cm. — (Security and the environment)
   Includes bibliographical references and index.
   ISBN 978-0-313-38006-8 (hard copy : alk. paper)—ISBN 978-0-313-38007-5 (ebook)
1. Human beings—Effect of climate on. 2. Human security. 3. Climatic changes—Social
aspects. 4. Climatic changes—Political aspects. 5. Environmental management. I. Title.
   GF71.W43   2010
   304.2′5—dc22   2010000839

ISBN: 978-0-313-38006-8
EISBN: 978-0-313-38007-5

14   13   12   11   10   1   2   3   4   5

This book is also available on the World Wide Web as an eBook.
Visit www.abc-clio.com for details.

Praeger
An Imprint of ABC-CLIO, LLC

ABC-CLIO, LLC
130 Cremona Drive, P.O. Box 1911
Santa Barbara, California 93116-1911

This book is printed on acid-free paper ∞

Manufactured in the United States of America

To my parents,
Helga and Heinz Webersik

CONTENTS

ILLUSTRATIONS

FIGURES

TABLES

SERIES FOREWORD

Our key focus in the Praeger Security and the Environment series is to explore the interstices between environmental, geopolitical, and security impacts in the twenty-first century. To those intimately involved with these issues, their immediacy and importance are obvious. What is not obvious to many—including those involved in making decisions that affect our collective future—is how these three critical issues are in constant conflict and frequently clash. Today, more than ever before in history, intersecting environmental, political, and security issues impact our lives and the lives of those who are to come.

In examining the complex interdependence of these three impact effects, the study of security, geopolitical, and environmental issues should recognize several distinct and pragmatic truths: One, international organizations today are established for and focus best on security issues. Thus, although it remains difficult to address environmental threats, challenges, and vulnerabilities for these organizations, it makes imminently better sense to reform what we have rather than constantly invent the "new" organization that may be no better equipped to handle current and future challenges. Secondly, the introduction of new protocols must continue to be created, worked into signature, and managed under the leadership of states through international organizations and cooperative regimes. Finally, and incorporating the reality of these previous truths, we should honestly recognize that environmental challenges can best be presented in terms that relate to security issues. To that end, it is sensible to depict environmental challenges in language that is understandable to decision makers most familiar with security impacts and issues.

There is benefit and danger in this approach, of course. Not all security issues involve direct threats; some security issues, as with some political processes, are far more nuanced, more subtle, and less clearly evident. I would argue further—as I have been arguing for several decades now—that it remains a tragic mistake to couch all security issues in terms of threat. To the contrary, what I term "entangled vulnerabilities"—population growth; disease; climate change; scarcity of water and other natural resources; decline in food production, access, and availability; soil erosion and desertification; urbanization and pollution; and the lack of effective warning systems—can have a far more devastating impact effect if such issues are ignored and left unchecked over time. In the worst possible outcomes, vulnerabilities left unchecked over time *will* manifest themselves as threats.

Environmental security thus emphasizes the sustained viability of the ecosystem, while recognizing that the ecosystem itself is perhaps the ultimate weapon of mass destruction. In 1556 in Shensi province, for example, tectonic plates shifted and by the time they settled back into place, 800,000 Chinese were dead. Roughly 73,500 years ago, a volcanic eruption in what is today Sumatra was so violent that ash circled the earth for several years, photosynthesis essentially stopped, and the precursors to what is today the human race amounted to only several thousand survivors worldwide. The earth itself, there can be little doubt, is the ultimate weapon of mass destruction. Yet, viewed through a different lens, mankind itself is the ultimate threat to the earth and the earth's ecosystem.

In *Climate Change and Security: A Gathering Storm of Global Challenges*, Christian Webersik provides us with a convincing argument that climate change's human effect has already begun. He does so with careful, precise arguments that provide us with not only a primer on climate change but with considerations that are practical and necessary for the policy maker as well as the concerned citizen wishing to be better informed about this (often needlessly) enormous debate. He does so in decidedly "nonalarmist" terms; rather, he carefully frames his argument—and wisely so— in terms of human security and the migration, resource scarcity, natural disaster, and conflict impacts we are seeing in some of the already most fragile regions of the world. Equally, and to his great credit, he provides us with a new environmental security agenda as well—detailing the limits and the promise of mitigation, adaptation, and consequence results that we are beginning to see. *Climate Change and Security* is an important and necessary contribution.

Three decades ago, the environmentalist Norman Myers wrote that national security is about far more than fighting forces and weaponry. National security must also include issues of environment and environmental impact—from watersheds to climate impact—and these factors must figure in the minds of military experts and political leaders. Myers's words today remain as prophetic, and deadly accurate, as ever.

Environmentalists sometimes predict an apocalypse is coming: The earth will heat up like a greenhouse. We will run out of energy. Overpopulation will lead to starvation and war. Nuclear winter will devastate all

organic life. We have, of course, grown de-sensitized to many such prophe-
cies of doom. Webersik argues, nonetheless, that the time and the need for
strategic planning and strategic action are more pressing now than ever.

Webersik shows us the global environmental challenges before us. If we
do not heed the warning signs, then we imperil ourselves and our future.
With meticulous detail and yet an immensely readable argument, he illustrates how and why we have entered a new era—one where we can no
longer afford to be oblivious.

P. H. Liotta
Executive Director
Pell Center for International Relations and Public Policy
Salve Regina University
Newport, Rhode Island

ACKNOWLEDGMENTS

Many people have contributed to this work and supported me. In writing this book, I benefited from countless acts of support, wise teaching, and guidance.

I begin with my family, my parents Helga and Heinz Webersik, my sister Annette Rappl, and my aunt Inge Breit, for their loving support throughout. I acknowledge with gratitude the great debts I owe to my loving wife, Joanna Szeniszewska, for her understanding and support in this endeavor.

I am especially grateful to my series editor, Peter H. Liotta, for his clear professional guidance and instrumental help with the manuscript. The same applies to Robert Hutchinson, senior editor at Praeger Security International, who guided this project from the beginning through the final stages. The final product was made possible through Nicole Azze, production coordinator, and Rachel Neal, publicity assistant at ABC-CLIO. Katherine M. Grier, project manager at Cadmus Communications, guided me though the production process, and her assistant, Holly Collins, was instrumental at the proof stage. I am grateful to all three of them. Special thanks go to my editor Jo-Ann Parks who read and edited the entire manuscript several times with great rigor and professionalism.

I would like to thank all staff at the Centre for Development Studies, University of Agder, who share an appreciation for applied interdisciplinary research, for their support and guidance, particularly the Centre's director, Professor Arne Olav Øyhus.

My thanks also go to friends and colleagues who shared their experiences with me and generously gave me their time, especially at the United Nations University (UNU). Here, I would especially like to mention Miguel Esteban, currently Assistant Professor at Waseda University, who developed the methodology I presented in connection with Chapter 3, and for his pivotal role in the review process. Vesselin Popovski, Senior Academic Officer at the

UNU Institute for Sustainability and Peace was a keen supporter of my work. At UNU-IAS, my thanks go to W. Bradnee Chambers, Alexandros Gasparatos, Per Stromberg, and Miguel Chacon Veliz. I also would like to acknowledge Dexter Thompson-Pomeroy for his intellectually stimulating input, in particular for Chapter 5. My host professor at the Tokyo Institute of Technology Norichika Kanie and his assistant Yuko Ura gave me the necessary academic freedom and guidance to do the bulk of the manuscript.

Further, I would like to acknowledge Jens Weinmann at the European School of Management and Technology who was instrumental in the review process, providing me with numerous useful comments. I also would like to thank Jonathan Philipsborn for his invaluable contribution, especially for the Sudan section in Chapter 2. My thanks also go to Manish Thapa, Robert McNamara Visiting Fellow at the Department of Peace & Conflict Research, Uppsala University, who facilitated my research in Nepal, reflected in chapter 2.

The idea to write this book is an outcome of my tenure as postdoctoral fellow at Columbia University's Earth Institute and the Center for International Earth Science Information Network (CIESIN). The intellectual and stimulating exchange with many friends, colleagues, and fellows created the foundation of this book. I am especially grateful to Bob Chen, Alexander de Sherbinin, Christopher Doll, Susan Doll, Meredith Golden, Christian Klose, Marc Levy, Franco Montalto, John Mutter, and Jeffrey Sachs.

My mentors at Oxford University and the University of East Anglia, Professor William Beinart, Dr. David Turton and Professor Timothy O'Riordan, lay the foundation on which I could build a solid academic career, and many of the ideas developed further in this book stem from the work I did under their incomparable supervision and invaluable guidance. I am very grateful to them.

Earlier versions of some of the material of the six chapters have appeared as articles in *Climatic Change, Natural Hazards, Sustainability Science, and Sustainable Development.* The case study material in chapter 2 was also used for a UNU-IAS factsheet series. I presented some sections of this work to the annual meetings of the American Political Science Association in Chicago and Toronto, to the annual meetings of the International Studies Association in San Francisco and New York, and the Seventh Open Meeting on the Human Dimensions of Global Environmental Change in Bonn, Germany.

I would like to thank the Japan Society for the Promotion of Science (JSPS); without its generous funding during my tenure as JSPS-UNU fellow, this research would not have been possible. I also would like to acknowledge the Centre for Development Studies and the faculty of Economics and Social Sciences at the University of Agder for providing financial support for this work.

INTRODUCTION

Ever since Vice President Al Gore and the Intergovernmental Panel on Climate Change were awarded the Nobel Peace Prize in 2007, scientists, the public, and the media acknowledged the connection between climate change and security. Since then, a number of policy reports, academic articles, and books emerged—some fairly dramatic, others more reasoned—drawing international attention to climate change impacts on humans and our environment, complementing the discussion on the human factor causing climate change.

Heat waves in France killed thousands of people in 2003; the Katrina flooding of New Orleans left more than 1,800 people dead in 2005; tropical cyclone Nargis in Myanmar made thousands of people homeless in 2008; the rapid melting of the summer sea ice in the Arctic opened up new shipping routes in 2007. These are all events that have something in common—as many analysts believe, they are not isolated events but a sequence of events indicative of climate change. If unchecked, climate change is very likely to have catastrophic consequences for ecosystems and humans, the rich and the poor, the young and the elderly, in all parts of the world.

But what does the term *security* mean in the context of climate change? Traditionally, security referred to a political or military threat to national sovereignty. Since the end of the Cold War and even before, scholars and policy makers alike have broadened the conventional definition of security to include the growing impacts of environmental stresses on human security and international security, and more recently those associated with climate change. As early as 1995, Levy associated security with climate change impacts.[1] He argued that human health is the only hazard that by itself can pose a security risk, especially the spread of malaria and other insect-borne diseases. The combination of other climate change impacts, such as sea-level rise, loss of wetlands, and agricultural productivity loss due to erosion, would lead to welfare losses including the U.S. domestic economy, making social upheaval probable. In fact, as demonstrated later in the book, economic shocks caused by an abrupt precipitation change can elevate the risk of internal conflict in countries with climate-sensitive economies. These outcomes, however, are exceptions rather than the rule. Most climate change impacts will have negative impacts for humans, threatening human well-being, rather than national or even international security.

Ever since the United Nations Development Programme (UNDP) defined human security as "freedom from want" and "freedom from fear" in its 1994 *Human Development Report*, the concept gained momentum.[2] As in the UNDP report, I define the term *human security* fairly broadly, as protection from any type of threat (both chronic and sudden) disrupting daily life that undermines human well-being. Of course, this very broad concept has been criticized in terms of utility for both policy makers and academia. The concept is accurate but it lacks precision. In Paris's words, human security tends to be "extraordinarily expansive and vague, encompassing everything from physical security to psychological well-being, which provides policy makers with little guidance in the prioritization of competing policy goals and academics little sense of what, exactly, is to be studied."[3]

By expanding the term *security to human security*, the focus shifts from national sovereignty to human well-being. This emerging paradigm focuses on the individual rather than the state as the main referent for security. This approach highlights the importance of people's vulnerability and resilience—how they are affected by environmental change as well as the way they react to it. This stands in contrast to the "environmental conflict" approach, as Detraz and Betsill rightly observe, which highlights the possibility of humans engaging in violent conflict because of a shrinking resource base and population pressures.[4] By adopting this approach, climate change becomes a threat to national and international security, because it can drive people into armed conflict as their natural resource stock diminishes. What is needed here is a careful analysis of under what circumstances climate change induces conflict and when it affects human well-being.

Leaving aside climate change impacts, there are other good reasons to adopt a broader definition of security. Wars, which threaten national security, have multiple outcomes, the most serious being mortality.[5] Today, the impact of war goes far beyond battle-related deaths. Noncombat–related issues, such as deteriorating living conditions, disease, lack of medical provisions, and food insecurity affect civilians and soldiers alike. In addition, wars destroy property, assets, and infrastructure. Moreover, most military threats come from within countries; civil wars are the most common type of war in the twenty-first century, often affecting only parts of a country. Sudan and Somalia are good examples.

In recognition of this trend, the intelligence community has contributed to the discussion on climate change and security. The U.S. National Intelligence Council published two reports on the topic, highlighting a range of potential climate change–induced threats to the United States.[6] In a statement, Thomas Finger, Chairman of the National Intelligence Council, argues that it is less likely to see state failure triggered by climate change but there is potential that climate change threatens domestic stability with intrastate conflict, possibly over scarce water resources.[7] A follow-up report published a few months later confirms Finger's statement that climate change is unlikely to cause interstate war, while pointing out that cooperation over water resources will be increasingly difficult.[8] What the reports miss, according to Dabelko, is the fact that more and more unintentional conflicts, some of which have violent outcomes, take place in local settings connected to livelihoods and access and control over natural resource use.[9] Certainly, a state's military apparatus is an inappropriate answer to climate change threats. On the contrary, the military is often responsible for great environmental damage and pollution.[10]

Taking climate change seriously requires us to reconsider the conventional definition of security. Compared to a state-centered or realist approach to security, there is another important difference—apart from the type of impacts: Climate change is self-imposed and we are both responsible for and affected by the impacts. There is no clear victim–perpetrator relationship. This, of course, is not entirely true because many societies contribute only marginally to climate change but can expect the most severe consequences: the destruction of homes, a higher disease burden, and the loss of life.

In contrast to a "climate conflict" approach considering intentional conflict as an outcome of climate change impacts, I treat climate-related impacts, such as human migration and displacement caused by natural hazards, resource scarcity affecting food security, or sea-level rise as a human security issue. I deliberately avoid an alarmist and sensationalist approach to climate change outcomes—equating climate-induced sea-level rise with massive population movements and associated armed conflict—because this opens the doors to potential reactionary policies. Governments could block "environmental refugees" while ignoring the underlying political causes of forced migration.[11] Alternatively, governments may escape from their responsibility to resolve conflicts by blaming global climate change for human suffering. In this respect, it may be politically misleading to label conflicts as in Darfur as "climate change conflicts." In a *Washington Post* opinion column, United Nations (UN) Secretary-General Ban Ki-moon wrote that climate change is one of the triggers of the conflict in Darfur, claiming that other parts of the world will face similar problems in the future: "Amid the diverse social and political causes, the Darfur conflict began as an ecological crisis, arising at least in part from climate change."[12] This can lead to a deterministic view that conflict is inevitable and beyond human control. After all, linkages between conflict and climate change are extremely complex and multidimensional.[13] This is perhaps the best argument as to why it is difficult to consider climate change as a driver of armed conflict, for the simple reason that we know little about what causes conflict in the first place.[14] We need to rethink the term *security* and possibly even abandon it altogether, as Dalby suggests in his book *Security and Environmental Change*: "Security is about making things, notably our consumer society, stay the same, it may in fact be part of the problem, rather than a way of thinking that is helpful in dealing with the future."[15]

Climate change impacts on human interactions are not new. For thousands of years, people have altered their environment. Historically, there are examples of climate change that triggered severe and often violent socioeconomic change, such as the great nineteenth-century droughts in China, Brazil, and India. There is evidence that environmental degradation and the depletion of natural resources has led to the collapse of entire societies. Diamond's story about Easter Island's historical trajectory is a convincing example.[16] He argues that Easter Island's sophisticated societies—able to erect huge stone images—almost disappeared due to an overexploitation of a limited resource base. But today, humans are not only altering their environment but the entire earth system with adverse socioeconomic (human) security implications, such as drought-induced famines and humanitarian crises caused by tropical cyclones.

Global implementation of innovative technological, legal, and international governance solutions are required if the human species is going to successfully adapt to climate change by minimizing local, regional, and international conflict and human insecurity. If climate change continues unabated, the costs of adaptation will be enormous, and consequently, if the loss of national welfare is greater than a society considers tolerable, conflict may arise.[17] By contrast, the costs of mitigation are relatively small. The Stern report on the economics of climate change confirms this and estimates that the costs of inaction will be much higher than the mitigation costs. In more detail, the report estimates the mitigation costs at 1 percent of annual global gross domestic product (GDP).[18]

In 2007, the United Nations Security Council held its first session on climate change and security.[19] The permanent members of the Security Council are the world's biggest polluters, yet typically they are not the nations that will be the most affected by climate change. An exception is China, with the largest total greenhouse gas emissions worldwide but relatively low per capita emissions over the past hundred years. Still, the Security Council needs China's cooperation to deal with climate change. In 2007, China's annual carbon dioxide emissions accounted for 20.6 percent of the world total compared to 19.3 percent from the United States.[20]

Africa, only represented by a three-member elected bloc on the Security Council, is widely viewed as the world's most vulnerable region to climate change yet the least able to adapt and with the greatest potential for endemic conflict.

We can also expect looming conflicts over climate change mitigation and adaptation strategies. Drastic measures to reduce emissions are likely to lead to the proliferation of nuclear energy, problematic in countries that have unstable political regimes as in Iran or Pakistan. The greater reliance on biofuel producers such as Brazil may invoke geopolitical considerations. Food security has already led to protectionist measures. Countries scarce in renewables but rich in fossils are starting to invest in farmland overseas.[21] Geoengineering solutions such as loading aerosols into the atmosphere can change hydrological cycles, leading to increased evaporation and hence to droughts. These are just a few examples of the unintended consequences of climate change mitigation and adaptation that require political solutions.

The future involvement of the UN Security Council remains uncertain; what is certain is the additional stress climate change will put on societies, many of them already subject to extreme poverty, internal military threats, inequality, and a high disease burden. Ultimately, the planet does not depend on us; we need the planet to sustain our livelihoods and to provide a life in peace and economic prosperity for generations to come.

The remainder of this volume is divided into six chapters. Chapter 1 presents the basic science of climate change as pertinent to the security topics treated in this work. It describes the historical and contemporary socioeconomic impacts induced by climate change, and the literature on environmental security. The second part of Chapter 1 examines adverse and positive feedbacks on human and international security. Adverse security consequences of contemporary climate change include biodiversity loss, water scarcity, droughts and famines, greater disease burdens, migration, and progressive submergence of densely populated coastal areas and entire small island states. Some positive socioeconomic consequences include the opening of new shipping routes in the Arctic and improved agricultural production in the northern hemisphere.

Chapter 2 critically surveys the impact of resource scarcity on security. Being more dependent on renewable resources, poor countries are proportionately more vulnerable to environmental stress than rich countries. Climatic changes such as droughts put more pressures on resources and hence increase the vulnerability and poverty of rapidly growing populations in developing countries. Poverty in turn has socioeconomic effects, such as wider socioeconomic inequalities and lower costs to recruit rebel soldiers, which can fuel political instability and can culminate in political violence with potential spillover effects to other countries.

Chapter 3 discusses natural disasters and their implications for human security. Progressively warming oceans are leading to more intense and potentially more frequent meteorological natural disasters such as tropical cyclones, storm surges, floods, and droughts, often with catastrophic effects on human and, in a few cases, international security.

Chapter 4 considers migration as an outcome of climate change. Already entire low-lying small island states or coastal states such as Kiribati and the Gambia are at the risk of being submerged. Sea-level rise and sudden natural disasters will compel whole regions with growing populations to resettle temporarily or permanently, raising pressure on land and resources. The risk of human insecurity will increase as sea-level rise quickens, making social adjustments to the dislocation less gradual.

Chapter 5 looks at the ripple effects of climate change mitigation. Efforts to mitigate the consequences of climate change will have unintended social costs that will increase the risk of new conflicts. For example, as increasingly larger areas of arable land are given over to land-extensive renewable energy technologies such as biofuel, global agricultural production will decline and food prices will be driven up. Similarly, when nondemocratic governments such as Iran embrace nuclear energy as a substitute for fossil fuels, international insecurity is likely to increase.

Chapter 6 sets out to define a new environmental security agenda for the twenty-first century, surveying ways of climate change adaptation and early-warning systems, and concluding with an outlook on the legal and institutional mechanisms to deal with climate change in the future. The slow pace of climate change impacts will require an intergenerational contract. This is a challenge for all governments that need to address the issue of equity and justice while minimizing the impact of climate change on the most vulnerable—the poor, the elderly, and the young.

ABBREVIATIONS

ACLED	Armed Conflict Location and Event Data
ACTS	African Centre for Technology Studies
AR4	*Fourth Assessment Report of the International Panel on Climate Change*
CBD COP 9	9th Conference of Parties to the Convention on Biological Diversity
CCS	Carbon Capture and Storage
CEWARN	Conflict Early Warning and Response Network
CHRR	Center for Hazards and Risk Research
CIA	U.S. Central Intelligence Agency
CIESIN	Center for International Earth Science Information Network
CO_2	Carbon Dioxide
CPA	Comprehensive Peace Agreement
CRED	Centre for Research on the Epidemiology of Disasters
EM-DAT	Emergency Events Database
ENCOP	Environment and Conflicts Project
ENMOD	Convention on the Prohibition of Military or Any Other Hostile Use of Environmental Modification Techniques
FAO	Food and Agricultural Organisation
GDP	Gross Domestic Product
GHCN	Global Historical Climatology Network
GIS	Geographic Information Systems
GNP	Gross National Product
GPCC	Global Precipitation Climatology Center
GPCP	Global Precipitation Climatology Project
IAEA	International Atomic Energy Agency
ICBM	Intercontinental Ballistic Missile

ICIMOD	International Centre for Integrated Mountain Development
IDPs	Internally Displaced Persons
IPCC	Intergovernmental Panel on Climate Change
IRI	International Research Institute for Climate and Society (formerly International Research Institute for Climate Prediction)
ITTO	International Tropical Timber Organization
JSPS	Japan Society for the Promotion of Science
LDCs	Least-Developed Countries
N_2O	Nitrous Oxide
NAPA	National Adaptation Programme of Action
NASA	National Aeronautics and Space Administration
NATO	North Atlantic Treaty Organization
NDVI	Normalized Difference Vegetation Index
NOAA	National Oceanic and Atmospheric Administration
ODA	Overseas Development Assistance
PIK	Potsdam Institute for Climate Impact Research
ppm	Parts per Million
PPP	Purchasing Power Parity
PRIO	International Peace Research Institute, Oslo
REDD	Reducing Emissions from Deforestation and Forest Degradation
TAR	*Third Assessment Report of the International Panel on Climate Change*
TERI	The Energy and Resources Institute (formerly Tata Energy Research Institute)
UCDP	Uppsala Conflict Data Program
UN	United Nations
UNEP	United Nations Environment Programme
UNEP/GRID	United Nations Environment Programme / Global Resource Information Database
UNFCCC	United Nations Framework Convention on Climate Change
UNHCR	United Nations High Commissioner for Refugees
UNOCHA	UN Office for the Coordination of Humanitarian Affairs
UNU	United Nations University
UNU-IAS	United Nations University Institute for Advanced Studies
UNU-EHS	United Nations University Institute for Environment and Human Security
USAID	United States Agency for International Development
WMO	World Meteorological Organization

CHAPTER ONE

Impact of Climate Change on Security

> There is still time to avoid the worst impacts of climate change, if we act now and act internationally.
>
> Lord Nicholas Stern, October 30, 2006

A PRIMER ON THE SCIENCE OF CLIMATE CHANGE

Climate change is one of the greatest challenges of the twenty-first century. The challenge lies in the scale of the problem, its global origins, and its long-term impact. Inaction today will affect future generations. There is hope that technologies will enable us to mitigate climate change at a modest cost, and at much lower costs than those of inaction. But markets alone are not going to solve this challenge.[1] What we need are incentives, new technologies, and most importantly, an international agreement in combination with voluntary individual as well as collective action to reduce greenhouse gas emissions, mainly carbon dioxide (CO_2) drastically. The United States, Europe, Japan, Canada, and Australia are responsible for most of today's cumulative CO_2 concentrations in the atmosphere. If unabated, our climate will soon reach critical tipping points, with the loss of all summer sea ice in the Arctic together with the disintegration of the West Antarctic and the Greenland ice sheet, subsequently leading to an unstoppable sea-level rise. Other outcomes are biodiversity loss at a fast rate, more heat waves and stronger storms, and a changing hydrologic cycle with more floods, intense rains, droughts, and forest fires.[2]

The Earth's climate has been changing for millions of years; what is new is the human change shaping climate change. The remainder of this chapter presents the basic science of global climate change as relevant to the security issues examined in this book. The Intergovernmental Panel on Climate Change (IPCC) offers the following fairly conservative definition of climate change:

Climate change . . . refers to a change in the state of the climate that can
be identified (e.g., using statistical tests) by changes in the mean and/or the
variability of its properties, and that persists for an extended period, typically
decades or longer. It refers to any change in climate over time, whether due
to natural variability or as a result of human activity.[3]

Anthropogenic or human-induced climate change has already become a
reality. Figure 1.1 shows that our planet has become warmer over the past
decades. We can already observe some of the predicted outcomes of climate change: 11 of the years from 1995 to 2006 were among the 12 warm-
est years since the recording of global surface temperature in 1850.[4]
Within the past 100 years (1906 to 2005) our climate became 0.74°
Celsius warmer.[5] In addition, sea-level rise has occurred over the past

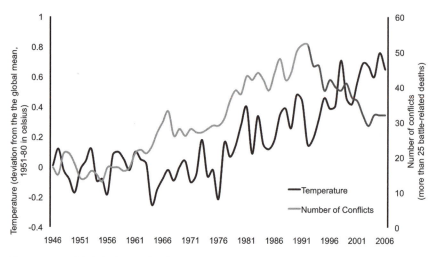

**Figure 1.1 Global Annual Mean Temperature and Armed Conflicts, 1946–
2007.** Both increased steadily until the early 1990s, leading to the (largely unsubstantiated) assumption that higher temperatures mean more conflicts. This trend
changed with the fall of the Berlin Wall, when the number of conflicts started to
decrease, a fact often overlooked in security studies.

Sources: Temperature data: Global Historical Climatology Network (GHCN),
1880–3/2008 (meteorological stations only), Climate Services and Monitoring
Division, NOAA/National Climatic Data Center, Asheville, NC (best estimate
for absolute global mean).

Conflict data: Lotta Harbom, Halvard Buhaug, Joachim Carlsen, and Håvard
Strand, "UCDP/PRIO Armed Conflict Dataset, Codebook, Version 4-2007" (Upp-
sala, Oslo: Uppsala Conflict Data Program [UCDP], International Peace Research
Institute, Oslo [PRIO], 2007); Nils Petter Gleditsch, Peter Wallensteen, Mikael
Eriksson, Margareta Sollenberg, and Håvard Strand, "Armed Conflict 1946–2001:
A New Dataset," *Journal of Peace Research* 39, no. 5 (2002): 615–37.

decade. Between 1993 and 2003, our oceans have risen by 3.1 mm. The majority—57 percent—of this rise is due to thermal expansion of the oceans, 28 percent is associated with the loss of ice caps and glaciers, and the melting of polar ice sheets contributes the rest.[6] In summary, over the past 50 years, the following trends have occurred:

- It is very likely that cold days, cold nights, and frosts have become less frequent over most land areas, while hot days and hot nights have become more frequent.[7]
- It is likely that heat waves have become more frequent over most land areas.
- It is likely that the frequency of heavy precipitation events (or proportion of total rainfall from heavy falls) has increased over most areas.
- It is likely that the incidence of extreme high sea level has increased at a broad range of sites worldwide since 1975.[8]

A combination of natural and human-induced forces contributes to climate change. These include changes in the atmospheric concentrations of greenhouse gas emissions (mainly CO_2, water vapor, methane, and nitrous oxide), aerosols, land cover, and solar radiation. These are all factors that can alter the energy balance of the Earth's climate system, thus influencing climate change. Solar radiation enters into the planet but greenhouse gases trap the heat by absorbing some of the reflected and outgoing infrared (long wave-length) radiation, thus generating a greenhouse effect.[9]

By assessing the level of human or anthropogenic influence on the current climate and compared to the *Third Assessment Report* (TAR) of the IPCC, scientists are now more confident (very likely in the *Fourth Assessment Report* compared to likely in the TAR) that the observed warming can be linked to an increase in anthropogenic greenhouse gas emissions. Whereas scientists can explain trends in global temperatures in the nineteenth century by changes in solar intensity and volcanic eruptions, they can only explain the temperature increase in the last 50 years by increasing greenhouse gas emissions.[10]

It is important to note that climate change is global in its causes and consequences. The main anthropogenic drivers are greenhouse gas emissions, mainly from fossil fuels, deforestation, forest degradation and land use change. There are other sources, such as commercial livestock production and industrial processes. The major greenhouse gases responsible for the anthropogenic forcing are CO_2, methane, and nitrous oxides. In terms of CO_2 emissions, the largest (56.6 percent) source is fossil fuels followed by forestry (including deforestation and forest degradation, and decay of biomass; 17.3 percent).[11] The largest share of global anthropogenic greenhouse gas emissions comes from the energy sector (25.9 percent), followed by industry (19.4 percent) and forestry (17.4 percent).[12]

Science has shown that a higher concentration of CO_2 and other greenhouse gases leads to a gradual warming of the atmosphere, with the subsequent warming of the oceans and the melting of glaciers and major ice sheets. Over the past decades, greenhouse gas concentrations in the atmosphere increased steadily. The current level of greenhouse gases in the atmosphere equals around 430 parts per million (ppm) CO_2. This is almost twice the preindustrial levels or stock of only 280 ppm.[13] Even if the flow of greenhouse gases would stabilize today, the stock in greenhouse gases would continue to rise and reach double preindustrial levels (around 1,750) by 2050, making climate change inevitable.[14]

The remainder of this chapter first examines the socioeconomic historical and contemporary impacts of climate change. It then turns to the literature on environmental security and its critics, and concludes with both adverse and positive feedbacks of climate change pertinent to human and international security. The priority here is the focus on the world's most vulnerable populations. This work does not aim at providing future estimates of people affected by climate change, nor does it attempt to specify what geographical regions will be future conflict hotspots. Rather, by relying on current scientific understanding of environmental processes, this book aims to shed light on how these processes affect human interactions.

SOCIOECONOMIC CONVULSIONS INDUCED BY CLIMATE CHANGE

Given the evidence of a changing climate and increasing greenhouse gas emissions, it is crucial to better understand the implications for us humans, who depend on the environment. The impact of climate on people is not new; it dates back to the very beginning of human kind. Likewise, humans have altered their environment ever since. It is only in the past 100 years that humans have started to alter the entire earth system. The following first discusses the historical socioeconomic changes induced by climatic change before turning to the anticipated contemporary socioeconomic implications of climate change.

Historical

Historically, climate change has triggered significant socioeconomic change, such as witnessed during the great nineteenth-century droughts in China, Brazil, and India.[15] For many centuries, humans both affected and were affected by environmental changes. Pollen diagrams, tree-ring analyses, and ice-core examinations are only a few of the scientific tools used to understand better our Earth's climate over the past 10,000 years. Even before that, humans significantly altered the environment. It is believed that in Australia some 45,000 to 50,000 years ago, human use of fire was responsible for land cover change and mega-faunal extinction.[16] Another example is the cultivation of rice in Asia dating back 10,000 to 14,000 years by substantially changing the landscape.

These are just a few preindustrial examples of how humans transformed the ecosystems they depend on. Historical societies often rely on climate-sensitive forms of agriculture, making them vulnerable to climate perturbations. Some of those societies, such as the people of Easter Island, the Mayan civilization, and the Khmer Empire in Southeast Asia, seem to have collapsed because of resource depletion and ecological degradation.[17] Another factor was the lack of adaptation, the lack of ability to cope with a changing environment.[18] There is no doubt that humans have shaped the planet's ecosystems for thousands of years, as well as being subject to environmental change as pointed out by Dearing:

> To argue that early human effects may be safely ignored in palaeoclimate reconstructions, or to assume that human activities were insignificant, runs counter to the voluminous amount of data worldwide for early and measurable responses to human activity. . . . Similarly, to ignore climate change as a potential element in affecting social behavior is to refute not just the growing number of regional and global palaeoclimate reconstructions that show that Holocene climate has been variable across a wide frequency range, but also the equally strong archaeological/anthropological evidence for climate being implicated in social change, and even collapse, on all the continents (except Antarctica)—especially in marginal agricultural environments.[19]

We also know from historical accounts that environmental perturbations did not affect people equally. The great famines in the late-nineteenth century, which killed no fewer than 30 million people, occurred in colonial Brazil, Africa, India, and China. This human calamity went almost unnoticed by the colonial powers in Europe, the United States, and Japan.[20] There is often an interaction between climatic and economic processes. Many of the people who died in the years 1877 to 1878 were already weak due to the world economic crisis commonly known as the "Great Depression," which began in 1873.[21] Even in times of famine, the challenge is often of an economic nature rather than one of food availability. Amartya Sen's entitlement approach explains situations in which food is actually moving out of the famine areas rather than into them.[22] A classic example, following Cecil Woodham-Smith's account, is the Irish potato famine of 1845 to 1851, in which one-fifth of the population died: "In the long troubled history of England and Ireland no issue has provoked so much anger or so embittered relations between the two countries as the indisputable fact that huge quantities of food were exported from Ireland to England throughout the period when people of Ireland were dying of starvation."[23]

The illustrations above provide only a snapshot of the many human-environment interactions long before human-induced climate change entered the scientific and policy discourse. However, the critical question remains—how can we learn from the past to direct future action?

Contemporary

When humans sent the first pictures of our planet from outer space, we started to comprehend the Earth as one globe, in all its beauty, as well as beginning to understand fully the planet's vulnerability. It was only some 15 years ago that scientists demonstrated that an increase in CO_2 emissions in the atmosphere correlates with a warming of global surface temperature (see Figure 1.2). What came to be known as "global warming" is, however, a much more complex phenomenon. Scientists today agree that the climate is changing; what is less clear is the level of certainty, speed, and direction of this change.

Much of the change we see today has its origin in the mid-eighteenth century. The industrial revolution marked a turning point in human history. The invention of the steam engine was the beginning of major changes in agriculture, mining, transportation, and manufacture. Industrialization and advances in science and technology improved human well-being for many. Urban areas grew, and modern commercial agriculture and better health care boosted Europe's populations. In parallel, humans started to consume more and more energy in industry, agriculture, and transportation. At the turn of the twentieth century, the main source of fuel was refined coal. Later, oil and gas gained importance to satisfy the growing economies' energy needs. Consequently, CO_2 emissions grew rapidly. With CO_2 concentrations reaching almost double the preindustrial level in 2010, human-induced climate change became a reality.

This work examines three major contemporary outcomes of global climate change: Natural resource scarcity, the increase in intensity and frequency of tropical cyclones, and climate-induced human migration. There are many more climate-induced effects, such as biodiversity loss or the melting of the polar ice sheets and glaciers; although important and linked to the former, I will deal with them in less detail.

Largely droughts, heat waves, and extreme rainfall with associated flooding shape resource scarcity. Droughts affect agricultural and water systems in semiarid regions, heat waves affect food production systems in mid-latitude areas, and extreme rainfall triggers landslides, erosion, and flooding of settlements.[24] Tropical cyclones cause flood and wind damage and often claim casualties, particularly in poor nations. They lead to economic losses, damage infrastructure (e.g., ports), and disrupt transport systems, mainly affecting coastal areas and settlements.[25] Climate change–induced human migration is triggered by tropical cyclones (short-term migration)

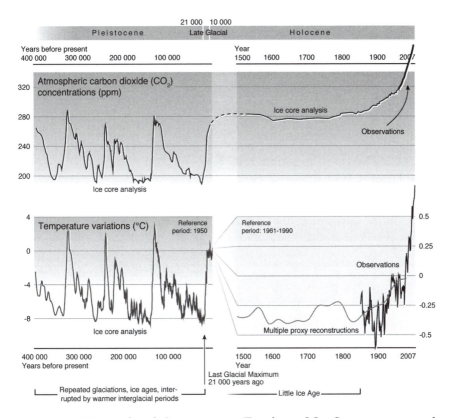

Figure 1.2 Historical and Contemporary Trends in CO₂ Concentrations and Temperature. Temperature averages have varied for the last 100,000 years, characterized by cycles of ice ages. Ice sheets covered large parts of this ancient world and low temperatures were the norm. At the same time, CO₂ concentrations, measured from tiny air bubbles trapped in the ice, were low. Since the Little Ice Age around 1850, temperatures have steadily increased. The top part of the curve derives from observations, and is referred to as the "Manua Loa Curve" or the "Keeling Curve."

Sources: Hugo Ahlenius, "Historical Trends in Carbon Dioxide Concentrations and Temperature, on a Geological and Recent Time Scale," UNEP/GRID-Arendal Maps and Graphics Library, 2007, http://maps.grida.no/go; J.-M. Barnola, D. Raynaud, C. Lorius, and N.I. Barkov, "Historical CO₂ Record from the Vostok Ice Core," in *Trends: A Compendium of Data on Global Change* (Oak Ridge, TN: Carbon Dioxide Information Analysis Center, Oak Ridge National Laboratory, 2007), http://www.cru.uea.ac.uk; D.M. Etheridge, L.P. Steele, R.L. Langenfelds, R.J. Francey, J.-M. Barnola, and V.I. Morgan, "Historical CO₂ Records from the Law Dome DE08, DE08-2, and DSS Ice Cores," in *Trends: A Compendium of Data on Global Change* (Oak Ridge, TN: Carbon Dioxide Information Analysis Center, Oak Ridge National Laboratory, 1998); C.D. Keeling and T.P. Whorf, "Atmospheric CO₂ Records from Sites in the SIO Air Sampling Network," in *Trends: A Compendium of Data on Global Change* (Oak Ridge, TN: Carbon Dioxide Information Analysis Center, Oak Ridge National Laboratory, 2005); M.E. Mann and P.D. Jones, "2,000 Year Hemispheric Multi-proxy Temperature Reconstructions" (Boulder, CO: IGBP PAGES/World Data Center for Paleoclimatology Data Contribution Series #2003–051. NOAA/NGDC Paleoclimatology Program, 2003).

or caused by sea-level rise (long-term migration). Migration can be short-term or permanent, voluntary, or forced. Largely poor countries with low adaptive capacity and small, low-lying island states may need to relocate their populations.

ENVIRONMENTAL STRESS FACTORS AND SECURITY

Before turning to the connections between climate change and security, it is worth considering the literature on environmental security. This debate is not new; it dates back to writings of Malthus, who published his "First Essay on the Principle of Population" in 1798.[26] Particularly since the end of the bipolar system, debate (in the literature known as neo-Malthusian) has heightened as to the contribution of environmental and demographic factors—environmental degradation, scarcity of renewable resources, natural hazards, and population pressures—to the outbreak of political violence.

Conventionally, the term *security* refers to the threat of physical force to national sovereignty. The Uppsala Conflict Data Program (UCDP) at the Department of Peace and Conflict Research, Uppsala University, and Centre for the Study of the Civil War at the International Peace Research Institute in Oslo (PRIO) have been recording conflicts since World War II. Figure 1.1 shows that the number of conflicts increased worldwide since the 1950s and peaked in the early 1990s at the end of the Cold War. Since then, the number of conflicts declined while global mean temperatures continued to rise. In the same vein, battle-related deaths also fell considerably to below 100,000 in the twenty-first century.[27] Considering population growth, the trend is even stronger.[28] Surely, we need to interpret these figures with caution because wars often lead to the disruption or breakdown of data-collecting institutions such as censuses, surveys, and statistical offices.[29]

Besides the overall decline in both conflict numbers and battle deaths, it is worth noting that the post–Cold War era is characterized by a decline in intrastate conflicts or conflict between states. This is due partly to the breakdown of the bipolar political system and partly to more effective peacekeeping and peacebuilding interventions from the international community.[30] There are also fewer new wars,[31] with recurring conflicts becoming more common. Consequently, one of the main risk factors to security is whether a country has a history of peace or conflict. Not surprisingly, 2005 and 2006 were the first two years with no new wars breaking out.[32] Most of the conflicts are concentrated in Africa and Asia, whereas during the Cold War, conflicts occurred in all parts of the world. What is more, most armed conflicts in the post–Cold War era affect the poorest and most vulnerable countries including Somalia, Sri Lanka, Sierra Leone, Chad, and Haiti, to name just a few. It is therefore important to redefine security to reflect these changes more accurately. Still, the number of armed conflicts is high, with 30 internal conflicts with more than 25 battle-related deaths taking place in 21 different countries in 2008. Two of those had more than 1,000 battle-related casualities, coded as major wars.[33]

Incorporating environmental factors into conflict analysis reaches back to the 1960s and 1970s.[34] Countries with rapidly growing populations, environmental degradation, and limits to food production were seen as a major threat to international security. By "securitizing" environmental concerns, researchers and environmentalists alike hoped to attract the attention of policy makers.[35] A range of academic work, largely using case studies, has developed since. Some are clearly provocative—for example, the often quoted and widely criticized article by Robert Kaplan on "The Coming Anarchy," where he argues that overpopulation coupled with environmental scarcity, crime, and disease are "rapidly destroying the social fabric of our planet."[36] He further argues that "West Africa is becoming the symbol of worldwide demographic, environmental, and societal stress."[37]

Other, more policy-oriented studies followed. In the United States, Vice President Al Gore commissioned the State Failure Task Force to examine the environmental, social, and economic causes of state failure. The research analyzed the forces that have caused instability in the post–Cold War era. With regard to environmental factors, the research findings argue that "environmental change does not appear to be *directly* linked to state failure,"[38] but that environmental stress affects the quality of life, using infant mortality rate as a measure. The authors identified infant mortality as a significant factor related to environmental degradation. In this sense, they made indirect linkages between environmental problems, social problems, and violence. Other studies were conducted by the Environment and Conflicts Project (ENCOP), which was jointly run by the Center for Security Studies and Conflict Research at the Swiss Federal Institute of Technology, Zurich, and the Swiss Peace Foundation in Bern. This project developed a typology of environmentally induced conflicts but did not explicitly link environment to violence.

A well-known scholar of this debate, Professor Thomas Homer-Dixon, undertook case study work and argues that environmental scarcity has social effects that can increase the risk of internal violence.[39] He distinguishes between three different types of scarcity: structural scarcity, demand-induced scarcity, and supply-induced scarcity. These three categories can be helpful in defining *scarcity*. In particular, structural scarcity goes beyond usually accepted definitions to include analysis of poverty and exclusion as well as, for example, discriminatory pricing structures of renewables. In this sense, the institutional setting with respect to control of natural resources is recognized as a vital aspect of discussion.

Environmental scarcity, he argues, influences agricultural and economic systems, causing migration and poverty.

Some studies have done more careful statistical research, such as the analysis by Wenche Hauge and Tanja Ellingsen, which found a significant impact from deforestation, soil degradation, and freshwater access on political violence.[40] Yet, conflicts are complex phenomena with many intervening variables that are difficult to model. Because relationships between ecological and political systems are complex, researchers often reject comparing case studies that do not display violence. For this reason, most of the case studies on environmental scarcity and conflict show some degree of violence, which is one of the main methodological weaknesses of this approach. They select cases on the value of the dependent variable, violent conflict, and exclude cases with a similar environmental context but no apparent outbreak of violence. A good example is Botswana, which has similar environmental features as, for instance, Somalia, but a history of peaceful development.

Another problem with more quantitative approaches—even based on large samples—is related to identifying the variables of greatest importance in proving the link between environmental factors and conflict. For example, even if infant mortality is a good signifier of overall material well-being, and is statistically linked to environmental stress and vulnerability, such a variable can only capture a small part of the contextual meaning of vulnerability or environmental stress, and it is difficult to relate it in any specific way to violence. A similar pitfall also applies to the statistical material presented in chapter 2. Though there is a correlation between low economic performance and rainfall, the link between low rainfall and elevated risk of civil war may mask other mechanisms that cause conflict.

Moreover, researchers have long seen population pressures on marginal lands and environmental resources as one of the main drivers of conflict. Some of the case studies anticipate environmental scarcity due to population growth leading to violent conflict. The authors use a "pie" metaphor to illustrate the causes of scarcity. The reduction of the resource base shrinks the pie, population growth increases the demand for resource usage per capita, and unequal income distribution divides the pie into pieces, some of which are too small to sustain a livelihood. For example, in Gaza in Palestine, environmental scarcity is caused by three factors: First, the depletion and degradation of water aquifers reducing the availability of water supply; second, population growth boosting demand; and third, inequitable water distribution between Palestinian and Israeli settlers.[41]

This neo-Malthusian strand of the literature has been criticized. Lipschutz argues that although researchers may have found a correlation between population growth and violence, this relationship does not justify an argument about causality.[42] Similarly, the linkage between environmental scarcity, population movement, or forced migration and conflict needs careful examination. It is important to incorporate environmental factors in the analysis of population displacement.[43] However, it is equally important to distinguish between the cause and effect, and to develop more complex explanations for the relationship between environmental scarcity, forced migration, and violent conflict.

For example, Henrik Urdal finds that countries with high population growth do not experience an elevated risk of armed conflict. This finding supports the hypothesis that more densely populated areas are forced to develop to overcome resource scarcity. Japan, contemporary India, and China are good examples. The study does not support the claim that the post–Cold War period represents a new era of insecurity. By contrast, conventional explanations of economic development, regime type (autocracies and democracies are the most stable), and geography account for much of the conflict we have seen in the past decades.[44] Also, according to the author, population movements do not seem to play a role in conflict onset.

When statistically controlling for trade, de Soysa makes the convincing argument that countries with high population densities have a higher propensity for conflict.[45] Urdal explains that land scarcity is a more pertinent issue, as the diversification of the economy through trade is restricted.[46] Humphreys and Richards make a similar argument that the lack of diversification of the largely agrarian economy of Sierra Leone was one of the factors that weakened the state and undermined economic development, leading to social unrest and grievances.[47] By comparing Pacific island states, one scholar finds that when migration is restricted, in combination with youth unemployment and environmental degradation, the risk of conflict increases.[48]

In other cases, the impact of people on their environment has been misinterpreted. Studies in Kenya have shown that density of population can actually reverse negative environmental trends. Tiffen, Mortimore, and Gichuki demonstrate the reversion of land degradation and environmental destruction in spite of increased population density and scarcity of fertile land.[49] The study of the Machakos District in Kenya from 1930 to 1990 explains how semiarid areas have been transformed "from an apparently misused and rapidly degrading latent 'desert' into a partially capitalised, still productive, and appreciating asset."[50] The study is a synthesis and interpretation of the physical and social development path in Machakos. The analysis focuses on long-term change in economic and environmental decision making by small farmers, in the context of population growth and environmental change, such as rainfall patterns, migration, and income diversification. This trend can also be observed in Niger, as shown in Figure 1.3, where a larger percentage of the country has become greener rather than drier in recent decades.

One explanation of this pattern of change derives from Boserup. She suggests that population growth fosters technological change, and argues

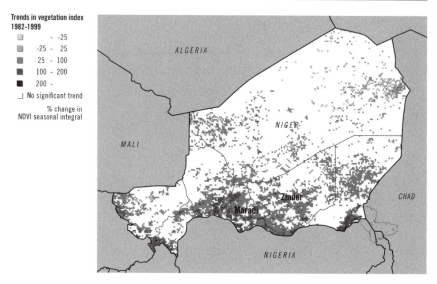

Figure 1.3 The Greening of Niger, 1982–1999. The Sahel, known for droughts and home to some of the poorest nations, has been subject to desertification and environmental degradation. The developments in Niger show a reverse of this trend, with a larger percentage of the country greening rather than drying.

Source: Hugo Ahlenius, "The Greening of Niger—Trends in Vegetation," UNEP/ GRID-Arendal Maps and Graphics Library, 2008, http://maps.grida.no/go; L. Olsson et al., "A Recent Greening of the Sahel—Trends, Patterns and Potential Causes," *Journal of Arid Environments* 63, no. 3 (2005): 556–66.

that scarcity of agricultural land induced by increased population density requires intensification of technologies already known that eventually lead to increased output per hectare.[51] Technological change, however, is only one of many factors determining environmental recovery in spite of Malthusian concerns. Tiffen and colleagues present three additional alternatives: Developing new land, which is usually of lower quality than the first settled; moving to a non-farm job, probably urban-based; and intensifying farming. The study suggests that income diversification is one of the factors stopping deforestation.

This leaves us with a dilemma. Broad statistical and comparative case studies that include environmental scarcity as part of their explanations of violence have some methodological weaknesses. Neo-Malthusian studies have been criticized for being deterministic and simply rather unspecified because they fail to identify those intervening variables that cause conflict.[52] Statistical studies on environmental security are often static without taking into consideration historical and structural causes, such as colonialism and globalization.[53]

A further, more general problem with the literature on environmental security is its tendency to focus on developing countries or non-Western countries. Only a few researchers from developing countries are vocal in the present debate. For instance, the former Minister for Defence of Rwanda, James K. Gasana, has written about natural resource scarcity and violence in Rwanda.[54] Further, the South African–based Institute for Security Studies has done research in the environmental security context, and the Nairobi-based African Centre for Technology Studies (ACTS) has undertaken studies examining ecological sources of conflict in sub-Saharan Africa.[55] The relatively small body of developing countries–based research suggests a bias in the focus of the literature. The non-Western approach is based on two assumptions, which are often implicit in the literature; first, that higher levels of general economic well-being and lower levels of scarcity in Western countries reduce the likelihood of environmentally related conflict there; and second, that Western countries have mechanisms for resolving environmental scarcity peacefully.

The field of environmental security has been of particular interest in the United States, and authors such as Mutz and Schoch explain this in relation to a shift in security policies of the U.S. government. They argue that in a context where the major justification for the use of force is no longer perceived threats to international security by the two opposing global hegemonic powers, national interests have become the legitimate justification for the use of force.[56] Environmental security is invoked, they suggest, whenever national interests are threatened by conflicts with their roots in environmental degradation or forced migration. In a similar way, Barnett argues that "the environment-conflict thesis is theoretically rather than empirically driven, and is both a product and legitimisation of the North's security agenda."[57] He comments that the United States publishes the majority of the literature, and states that the environmental security material is predominantly concerned with resources of economic value.

However, environmental issues seem to play a role in certain types of conflicts. Clearly, it is necessary to analyze the overall political, economic-historical framework, and the global linkages that make them significant. In his critical examination of Kaplan's "New Barbarism," Paul Richards convincingly makes just such a case with respect to Sierra Leone.[58] He does not find a correlation between deforestation and war. He argues that after the end of the Cold War, the availability of valuable resources for patrimonial redistribution declined. As a result, the state's control over the country's periphery weakened, giving rise to rebel violence. Young men did not join the rebellion because of opportunism but due to terror and coercion. He concludes that the war in Sierra Leone "is a product of this protracted, post-colonial, crisis of patrimonialism."[59] Abdullah supports this view and claims that in the post-colonial era there was a lack of an alternative political culture to the bankrupt All Peoples Congress, giving rise to the Revolutionary United Front in Sierra Leone.[60]

Another way forward is to explore more thoroughly the question raised in the discussion of the Norwegian sociologist and peace mediator Johan Galtung: that of equity and exclusion. In this regard, it is important to note that environmental scarcity is often less a problem of the absolute lack of resources than one of distribution of environmental resources. Studies concerned with environmental scarcity hardly consider the question of equal distribution of resources between the North and the South and among present and future generations.[61]

Distributional issues are as important as trends in the country's economic development. We know from studies of civil war onset that low economic growth (rather than the absolute level of wealth) elevates the risk of armed conflict.[62] In particular, in countries that have climate-sensitive economies, with a large agricultural sector, great differences in rainfall levels from one year to another can increase the risk of losing economic income.[63] Low economic growth leads to unemployment and decreasing levels of wealth. This in turn breeds antigovernment sentiments and grievances while undermining the government's legitimacy. By contrast, economic wealth measured in the gross domestic product (GDP) per capita or lower levels of infant mortality reduce the risk of conflict.

There are other reasons that countries experience conflicts—for example, a history of previous conflicts.[64] War is development in reverse, and many of the conflicts of the past decades are recurring conflicts between the same conflict parties.[65] The longer the peace lasts, the better the chances are that a country will remain peaceful. The nature of the war—whether or not it was very destructive—is also important. Destructive wars tend to be followed by more fragile periods of peace.[66] In addition, the nature of the peace agreements and the role of third parties influence outcomes. Outright victories tend to be more stable than negotiated settlements, and peacekeepers reduce the risk of conflict recurrence.[67] These are just a few factors that influence war outcomes.

Overall, environmental and demographic stress factors are important in shaping human interactions, but it is less likely that environmental degradation, resource scarcity, and population pressures alone act as a sufficient factor to trigger internal or even intrastate wars. Although the capturing of fertile land and livestock is a common characteristic of warfare in countries such as Rwanda and Somalia, I find the root causes of the conflict elsewhere.[68] Natural resources are important to all kinds of social interactions, and often local conflicts are over their distribution. There is a need for better data, improved methods, such as geographic information systems (GIS), and multilevel studies to understand better the feedback between people and their environment.

FEEDBACK OF CLIMATE CHANGE ON HUMAN AND INTERNATIONAL SECURITY

Although environmental security concerns are established topics in academia and policy debates alike, it is only relatively recently that climate change impacts have been included in this debate. A search for the term *conflict* in historical documents on climate change issues, ranging from the assessment reports of the Intergovernmental Panel on Climate Change to the Stern report, has revealed a somewhat limited use of this term. In only nine out of 23 documents was the word "conflict" found. In the historical documents, "conflict" was mentioned mainly in the context of water resources.

More recently, the focus—including the tone of this book—shifted toward climate change as a threat multiplier, phrased in terms of "environmental security." It is in this context that *in*security means increasing vulnerabilities and a limited capacity to adapt to climate change impacts. In Detraz and Betsill's analysis, security in the tradition of armed conflict was virtually absent.[69] Though mentioned more often than conflict, peace and security were referred to in only one case—a speech by former UN Secretary-General Kofi Annan: "Climate change is also a threat to peace and security. Changing patterns of rainfall, for example, can heighten competition for resources, setting in motion potentially destabilizing tensions and migrations, especially in fragile States or volatile regions. There is evidence that some of this is already occurring; more could well be in the offing."[70] A more recent and more dramatic statement, also included in Detraz and Betsill's analysis, comes from former UNFCCC Executive Secretary Yvo de Boer. He claimed in 2007—perhaps a bit prematurely—that, if no action was taken, "the consequences of climate change could plunge the world into conflict. In 2010, there could be as many as 50 million environmentally displaced persons as a result of climate change, desertification and deforestation. Competing for water, energy, and food can lead to ethnic rivalry and regional conflicts."[71]

Detraz and Betsill distinguished in their analysis between an "environmental security" and an "environmental conflict" discourse, and according to them, it is the former that has largely informed the climate change discourse despite Annan's and de Boer's alarmist claims. This finding is not surprising since there is little evidence that environmental change or climate change impacts translate directly into armed conflict. Under certain circumstances, however, when drastic environmental shocks lead to wealth deprivation and in countries with additional stress factors, such as poverty and inequality, environmental change can trigger violent outcomes. But what are the predicted climate change impacts that fall into such an "environmental security" discourse? The following lists the main climate change feedbacks I identified and included in the discussion that follows.

Adverse Feedbacks

There is a spectrum of negative and positive feedback of contemporary climate change on human and international security. Adverse consequences of present climate change include biodiversity loss, water scarcity, droughts, greater disease burdens, human migrations, and progressive submergence of populated coastal areas and the disappearance of entire low-lying small island states, such as the Maldives. Much of the following considers the contemporary socioeconomic impacts of climate change on human security (such as drought-induced famines and humanitarian crises caused by tropical cyclones) as well as—but to a lesser extent—international security (conflicts that are exacerbated by climate change with spill-over effects to other countries).

A number of recently published policy reports have made alarmist claims that environmental change will have an enormous impact on human security and implications for international security. A report published by Christian Aid estimates that up to 1 billion people will be forced to move between today and 2050, with serious implications for international security:[72] "The danger is that this new forced migration will fuel existing conflicts and generate new ones in the areas of the world—the poorest—where resources are most scarce."[73] In the same vein, a report signed by 11 retired U.S. generals and admirals argues that "climate change can act as a threat multiplier for instability in some of the most volatile regions of the world, and it presents significant national security challenges for the United States."[74] The Toronto-based scholar Homer-Dixon sees the need for the involvement of the United Nations Security Council to deal with the issue:

> Evidence is fast accumulating that, within our children's lifetimes, severe droughts, storms, and heat waves caused by climate change could rip apart societies from one side of the planet to the other. Climate stress may well represent a challenge to international security just as dangerous—and more intractable—than the arms race between the United States and the Soviet Union during the cold war or the proliferation of nuclear weapons among rogue states today.[75]

A report for the U.S. government formulates this "new threat" in even more drastic terms:

> All of [. . .] progressive behavior could collapse if carrying capacities everywhere were suddenly lowered drastically by abrupt climate change. Humanity would revert to its norm of constant battles for diminishing resources, which the battles themselves would further reduce even beyond the climatic effects. Once again warfare would define human life.[76]

Other reports—including the tone of this book—are more cautious, such as the IPCC's *Fourth Assessment Report* or the *Stern Report on the Economics of Climate Change.*[77] The focus of the analysis of the following is people-centered, employing concepts of social vulnerability and resilience. Along these lines, a report commissioned by the German government argues "that unabated climate change will increase human vulnerability, worsen poverty, and thus heighten societies' susceptibility to crises and conflicts."[78] Climate change and the discussion of security gained momentum within the last 15 years. Within this period, scientists started recording environmental changes such as sea-level rise or the increase in world surface temperature. In parallel, however, the frequency and intensity of armed conflict has significantly decreased worldwide (see Figure 1.1). In addition, the empirical foundation for claims made by alarmist reports is often weak or missing. To contribute to a more balanced debate, this work examines three adverse feedbacks of climate change—resource scarcity, natural hazard intensity and frequency increase, and human migration—and assesses their security implications.

In states with weak governance, youth unemployment, and low economic performance, sudden environmental shocks, such as droughts, can lead to resource scarcity and in turn to lower economic growth, which elevates the risk of political instability. This is particularly true in countries with a high dependency on agricultural production. The effect is even stronger in countries where the majority of agriculture is rain-fed, thus climate sensitive. Irrigation agriculture, by contrast, has the advantage to store water in times of scarcity. However, environmental change is only one of the many intervening factors that shape violent outcomes. In rich countries, heat waves are likely to have an impact on food security, as experienced in France in 2003. This shows that adverse feedback is not limited to the developing world. Still, the poorer regions are disproportionately more affected because their capacity to adapt is low while their dependency on climate-sensitive economies remains high.

The implications of an increase in intensity and frequency of natural hazards, largely tropical cyclones, floods, and droughts, on human security are clear: Storms, floods, and droughts have the potential to destroy people's food systems and infrastructure, and in the worst case, claim lives. Even in regions where human casualties are rare, as in Japan, tropical cyclones destroy assets and disrupt infrastructure, while lowering economic productivity.

Climate-induced human migration has a very limited potential for armed conflict but bears serious implications for government planning and policy. Already today, countries are struggling to assist internally displaced persons (IDPs). Recognizing "environmental refugees" and granting them special rights would push international agencies, such as the United Nations High Commissioner for Refugees, and receiving countries to their limits.[79] Certainly, there is a great potential for human migration, both voluntary and forced (in case of a sudden natural hazard like Hurricane Katrina), in particular in urban low-lying coastal zones.

Another set of security issues are so-called knock-on effects of climate mitigation. These are indirect, negative feedbacks of climate change mitigation. For instance, the substitution of coal with biofuels as promoted by key industrialized nations, such as the United States and Germany, has implications for global food security and biodiversity. Palm oil plantations in Indonesia compete for agricultural land and are often the main culprit of the destruction of pristine rainforests, home to endangered species like the orangutan.

Further, scientists and governments express great hope for geo-engineering to reduce emissions. These include fertilizing the oceans or capturing and storing CO_2 from industrial processes. However, many questions remain open: What are the environmental and long-term environmental risks of such geo-engineering endeavors? Can carbon capture and storage be a permanent solution while avoiding leakage? There is definitely a need for more research and the right policies to deal with these issues.

Another contested issue is the proliferation of the commercial use of nuclear energy as a way to substitute for fossil fuels. Nuclear power plants generate waste that can contaminate soil for generations to come. Nuclear reactor accidents, as occurred in 1986 at the Chernobyl Nuclear Power Plant in Ukraine, then part of the Soviet Union, can pose a health hazard for generations to come. Politically, as witnessed in North Korea, countries in the possession of commercial nuclear power plants may start producing uranium that they can use in nuclear warheads. The 2009 launch of a North Korean missile was widely condemned as posing a threat to international security.

Positive Feedbacks

New research demonstrates that all regions will experience adverse feedback. Nonetheless, there are some potential positive "externalities" of global climate change. Some of these positive socioeconomic consequences include new shipping routes in the Arctic and projected improved agricultural productivity in the northern hemisphere, largely in Alaska, Canada, the Scandinavian countries, and Russia. In terms of agriculture, CO_2 fertilization can offset some of the losses caused by water scarcity and heat. Some of the crops are, however, less susceptible to CO_2 fertilization. Researchers estimate that especially maize, the main crop in southern and western Africa, will experience net losses of 10 percent for a temperature rise of 3° to 4° Celsius above 1990 levels despite CO_2 fertilization.[80] Other crops, such as rice, the dominant crop in South and East Asia, may experience a 1 to 2 percent net increase for a temperature rise of 2° to 3° Celsius except for Central Asia and Africa.[81] Regarding new sea lanes, they have geopolitical implications and social costs by potentially creating environmental disasters for Arctic biota and indigenous peoples.[82]

CONCLUSION

As shown above, climate change impacts on human interactions are not new. Humans have altered their environment for thousands of years, and they have been subject to climatic changes shaping our societies. What is new is the scale and speed at which the current climate system is changing. Much of the observed warming can be contributed to human activities, and even if we manage to reduce emissions drastically, the stock of carbon dioxide will continue building up. Given the advances in climate modeling and forecasting, we have a relatively good picture of how increasing carbon stocks contribute to global warming, or how positive feedback loops can accelerate this process. By contrast, little is known about the socioeconomic security implications of climate change, an issue this book aims to address.

Studying interactions between humans and their environment is not new. Environmental security studies that tie the concept of "security" to "human security" rather than "armed conflict" emerged as early as the 1960s. Relatively new is the debate on climate change and security, a discussion that has gained momentum in recent years. The mention of climate change in connection with the Darfur conflict, several high-level statements, and alarmist reports from military think tanks that consider climate change as a new security threat have increased over the past years. I am more careful in my analysis, considering three distinct outcomes of climate change, resource scarcity, meteorological disasters, and migration and their negative impacts on humans. This approach stands in contrast to the "environmental conflict" approach that focuses on the likelihood of humans to engage in armed conflict as a result of degrading environment—although that is relevant for the following chapter.

CHAPTER TWO

Resource Scarcity and Security Implications

Addressing the issue of the environment in the context of conflict resolution, conflict prevention, peacekeeping, [and] peacebuilding becomes ever more important because we know from everything we have learned—and are learning every day—about climate change that one thing is for certain: The world is going to be under more stress.

Achim Steiner, Director, UN Environment Programme
May 31, 2009

INTRODUCTION

One of the likely effects of climate change is an increase in the scarcity of renewable resources in most regions of the world. Declining and changing rainfall patterns, a warmer climate, reduced freshwater availability, melting ice sheets and glaciers, and greater climate variability are all factors that scientists believe induce resource scarcity. The meaning of scarcity in this context is twofold: On the one hand, it refers to the actual effect on renewability of natural capital as measured by natural indicators of resilience and fragility. On the other hand, it refers to the relationship between declining availability of natural resources and local power relationships, property rights, entitlements, and prices, set against the ability to pay, and informational needs over conservation and redistribution. A gradual increase in resource scarcity could be offset by adaptation measures and coping mechanisms. However, the challenge is the increasing uncertainty inherent in the global climate system and the subsequent unpredictability of resource variability.[1]

Being more dependent on renewable resources, developing countries in the South are becoming proportionately more vulnerable to climate change, compared to industrialized countries. The Intergovernmental

Panel on Climate Change (IPCC) *Fourth Assessment Report* confirms this: The environmental impacts of climate change will be more severe in the low-income countries.[2] There are several reasons: One is the lower capacity of poor countries to adapt to climate change, another relates to geography, while others can be attributed to demographics or the structure of the economy. The economies of low-income developing countries are more vulnerable, because they are largely agrarian in nature. Households in poor countries spend proportionally more money on food. When food prices rise, African households will have proportionally less money to spend on education, health, housing, and food. In addition, climate shocks destroy household assets such as livestock. In many countries, animals mean collateral for credit, nutrition, a source of income, and security when crops are failing.[3] When livestock is lost due to drought, a household's future vulnerability increases.

But the adaptive capacity of poor countries to address the challenges of climate change remains limited due to low incomes and low institutional performance. Consequently, and as argued elsewhere, societies with weak adaptive capacities will be affected most.[4] A number of policy reports have underpinned this point, including the *Stern Report on the Economics of Climate Change* and a recent study by the German Advisory Council on Global Change.[5]

The chapter is divided into three parts. The first deals with three important aspects of climate change and their impact on resource scarcity: increasing temperatures, water stress, and climate variability, manifested in droughts and floods. The second part draws on statistical evidence based on the author's research about the impact of sudden drops in rainfall in sub-Saharan Africa on the propensity of political violence. The third and last part of this chapter illustrates three brief case studies, Somalia, Sudan, and Nepal. All three countries have climate-sensitive economies and a history of armed conflict. Here, climate change and conflict must be put in context, considering vulnerabilities and how it weakens the already fragile economy and social systems. Environmental change in this context acts as a "threat multiplier" linked to other stresses such as poverty, inequality, lawlessness, or the intervention of outsiders.[6]

CLIMATE CHANGE AND RESOURCE SCARCITY

One problem that is likely to affect both poor and industrialized countries is an increase in average mean temperature. Heat waves are one of the likely outcomes of a warmer climate. The record-high 2003 summer temperatures in Western Europe claimed more than 52,000 lives, with the most fatalities in France,[7] where the mean temperature was 3.6° Celsius above the long-term average. From August 1 through August 20, 2003, 14,800 people perished.[8] This increase in deaths followed the patterns of increase in temperatures.[9] In parallel, agricultural crop yields fell significantly due to heat stress and reduced freshwater availability. In France alone in 2003, maize production and fruit harvests fell by 30 and 25 percent, respectively.[10] In 1972, the former Soviet Union experienced an exceptionally hot summer, leading to a 13 percent decline in overall grain production.[11] The Soviet Union bought grain on the international market to compensate for the domestic losses instead of balancing shortfalls in fodder through culling of domestic livestock. As a result, developing countries, largely in Asia, feared political instability when wheat prices rose. Interestingly, in both cases, in 2003 in France and in 1972 in Russia, the precipitation deviations were not as large as the temperature deviations.[12] As this finding demonstrates, heat can cause severe damage to agricultural systems in the absence of a meteorological drought. Recent studies have underestimated this aspect of climate change.

Turning to the developing world, it is highly likely (the probability is greater than 90 percent) that by the end of the century the growing season temperatures will exceed the most extreme growing season temperatures measured between 1900 and 2006 for most of the tropics.[13] We know that higher mean temperatures signify reduced soil moisture and lead to an increased demand in water for irrigation. Already today, the heat in the Sahel can be so great that rain evaporates before it reaches the soil.[14] In addition, excess heat affects the grain development in a negative way. This will have devastating consequences for countries with a large agricultural sector and low irrigation and water-storage capacity. Apart from food security issues, research suggests an increase in civil war risk caused by temperature increase.

Burke and others project an additional 393,000 battle-related deaths increase or a roughly 54 percent increase in armed conflict in sub-Saharan Africa by 2030.[15] They argue that future temperatures over the next decades are more uniform whereas rainfall predictions vary widely, a caveat that needs to be considered in the following analysis on linking rainfall to conflict. However, there are some methodological pitfalls. The projections of Burke and others use a 20-year baseline from 1981 to 2002. Particularly, the conflict data varies considerably before and after this baseline, whereas temperatures increased steadily (see Figure 1.1). Also, other important intervening socioeconomic variables may change in the future, raising the question whether the model results are robust. This needs to be answered by future research.

Increasing temperatures will also lead to a melting of the freshwater resources in the Arctic, the Andes, and the Himalayas. All major river systems of the Indian subcontinent originate in the Himalayas. A warmer climate will first lead to an increase in water flow but in the long term, it will reduce the overall water flow, affecting some 500 million people downstream (see Table 2.6). By incorporating these effects, the IPCC *Fourth Assessment Report* calculates a 10 to 30 percent reduction in mean river runoff in dry regions and the dry tropics by 2050.[16] To make things worse and as shown in the Nepal case study, mean temperatures in the higher altitudes of the Himalayas are increasing at a faster pace than temperatures in the low-lying areas.

Resource scarcity is also a function of increasing demand for food. The current population growth of 1.2 percent puts more pressure on renewable resources, such as fresh water and arable land.[17] At the same time, the demand for biofuels makes food more expensive. Biofuel plants, such as cassava, oil palms, or maize competes with food crops and scarce arable land. In addition to an increase in demand, and as witnessed in Egypt in 2008, countries may introduce protectionist policies to safeguard domestic food supplies and to stabilize prices. This will create an artificial scarcity accompanied by more price hikes.

In addition, increasing climate variability will probably increase the scarcity of resources and vulnerability and poverty of growing populations in developing countries. Climate change is expected to increase the number of extreme weather events, such as droughts and floods.[18] This will especially affect sub-Saharan Africa. Compared to other parts of the world, Africa has the highest economic loss risk due to droughts (see Figure 6.1). The challenge is not only absolute scarcity of resources but climate variability. Climate variability influences economic performance. Research shows that high climate variability correlates with lowered economic performance. As a consequence, countries with high climate variability in sub-Saharan Africa tend to have lower per capita income.[19]

Most important, water stress will increase in most parts of Africa largely because of an increase in water demand.[20] As already mentioned, population growth and gradual industrialization are the main factors. In Africa, the overall population will grow to nearly 2 billion in 2050.[21] It is likely that climate change will exacerbate this condition. Experts estimate that 75 to 250 million and 350 to 600 million people in Africa will experience water stress by the 2020s and the 2050s, respectively.[22] Another source calculates that the proportion of the population facing water stress will increase to 65 percent in 2025.[23] Climate variability and water stress will also affect agricultural production (see Figure 3.5). One study forecasts that by 2100 net crop revenues could fall by 90 percent with small-scale farmers affected the most.[24]

Declining agricultural productivity can lead to low incomes that, in turn, have socioeconomic effects, such as wider socioeconomic inequalities and lower costs to recruit rebel soldiers, potentially fueling political instability and eventually culminating in violence. Jeffrey Sachs argues that there is a strategic significance to global economic inequalities.[25] Poor economic performance raises the risk of state failure, posing a potential threat to international security.

RESOURCE SCARCITY, DROUGHTS, AND SECURITY: EVIDENCE FROM AFRICA

The study of resource scarcity and civil war onset has become a major focus of attention in scholarly literature. The following argues that a negative change in rainfall growth (the difference in rainfall from year to year) in sub-Saharan countries is associated with higher risk of armed internal conflict. Societies that are dependent on renewable resources, particularly in the developing world, are more vulnerable to environmental stress such as erratic rainfall. In most sub-Saharan economies, the manufacturing and industrial sectors are in preliminary stages. Instead, agriculture constitutes a large percentage of overall national economic income. In some countries, such as the Central African Republic, Ethiopia, and the Democratic Republic of the Congo, the share of agriculture in the gross domestic product (GDP) in 2005 was as high as 54 percent, 48 percent, and 46 percent respectively.[26] Moreover, most agricultural production is based on rain-fed agriculture. In Africa, less than 7 percent of the overall agricultural production derives from irrigated lands.[27] As a result, changes in rainfall patterns can lead to drought, translating into increased vulnerability and hence poverty (see Figure 2.1). In turn, low incomes are associated with a higher risk of political instability.[28]

In theory, water scarcity induced by sudden rainfall shortages will lead to wealth deprivation, possibly increasing the likelihood of an armed rebellion. In this context, a sudden economic shock caused by failing rains frees labor, leading to an abundant supply of potential rebel soldiers with low opportunity costs. Consequently, in an environment with reduced economic returns caused by drought, the incentives to engage in a war economy increase. This argument is in line with the civil war literature that attempts to explain armed conflict by focusing on economic opportunity costs whether or not an individual is joining a rebellion.[29] As a working hypothesis, regions with a higher dependency on agricultural production will be more likely to experience civil conflict in a given drought year.

If rainfall is shaping economic conditions in sub-Saharan African countries, then changes in rainfall patterns indirectly induced by anthropogenic emissions can be associated with the outbreak of civil violence—a connection, however, that has not been tested rigorously. Only a few studies have provided evidence that climate change will lead to decreased rainfall, causing widespread poverty while lowering rebel recruitment costs.[30] Cullen Hendrix and Sarah Glaser have undertaken cross-country studies using interannual rainfall figures that confirm these effects.[31] Other authors arrived at similar conclusions using monthly rainfall data.[32] A recent article argues in favor of an "ecological origin of the Darfur crisis," blaming climate change for the death of thousands of people.[33] Along these lines, a

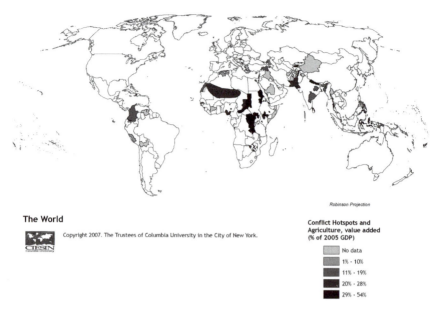

Robinson Projection

Conflict Hotspots and
Agriculture, value added
(% of 2005 GDP)

No data
1% - 10%
11% - 19%
20% - 28%
29% - 54%

Figure 2.1 Armed Conflict Hotspots and Agricultural Dependency. Regions
with armed conflict in 2006 had a high dependency on agriculture. As a hypothesis, climatic changes in these regions impact on economic growth, with repercussions for peace and security. Negative economic growth elevates the risk of
conflict; positive growth correlates with political stability.

Sources: The existence of conflicts was based on data from various sources, primarily the International Crisis Group's *CrisisWatch* newsletter of November 2006. The
spatial extent of conflicts was based on maps found on Web sites, including Relief-
Web.org and CrisisGroup.org, and in books, and which were in turn based on
information from 1999–2006. The Trustees of Columbia University in New York,
Center for International Earth Science Information Network (CIESIN), Palisades,
NY. Agricultural data (agriculture, value added [% of GDP]) were obtained from
the World Development Indicators 2006, the World Bank, Washington, DC,
http://web.worldbank.org.

recent study by the UN Environment Programme claims that "there is a
very strong link between land degradation, desertification and conflict in
Darfur."[34] Historical studies examined the impact of climate change on
warfare in Eastern China over the last millennium.[35] They find that cooling periods can be associated with a higher frequency of armed conflict.[36]
Other authors find little effect of environmental and demographic variables
on conflict onset, but state that local water scarcity, political marginalization,
high infant mortality, and proximity to international borders explain conflict
risks.[37]

It is evident that climate change will affect human well-being.[38] This is
particularly true for societies that are already vulnerable due to a high disease burden, poverty, and natural resource dependence. Climate change is

likely to increase pressures on these societies, but ethno-political exclusion remains important in explaining violence.

Linking Rainfall to Conflict

Although precipitation does not directly affect political stability, it is a robust predictor of economic performance, largely in countries dependent on smallholder agricultural production (see Figure 2.2). Research demonstrates that negative economic growth is associated with the onset of civil war.[39] Once conflict breaks out, other factors may lead to poor economic performance. Armed clashes destroy property, suspend industries, or disrupt farming systems. It is therefore difficult to observe the direction of the relationship of economic growth and civil war. To avoid this known problem of reversed causality, Professor Edward Miguel, director of the Center of Evaluation for Global Action at the University of California, used rainfall as an instrumental variable for economic growth.[40]

As suggested by the literature, countries with greater dependency on agriculture are more vulnerable to climatic variability and thus more likely to experience economic loss shaping political instability. This hypothesis shall be tested on a set of African countries with a high share of agriculture in their GDP. I therefore selected countries based on their dependency on agriculture. The selection criterion for the subsample was the percentage of agriculture adding to the gross national income (or GDP). Only countries whose share of agriculture is more than 30 percent were selected. Figure 2.1, based on a descriptive geographical analysis, shows that most countries that experienced armed conflict in 2006 were largely agrarian, especially in sub-Saharan Africa.

Modeling Change

To measure the dependent variable, armed conflict, I study internal armed conflicts and internationalized armed conflict.[41] Regarding the category internationalized armed conflict, it is crucial to select only those cases that occurred in the territory of the affected country, a caveat that was ignored by Miguel. At the time of writing, the Centre for the Study of Civil War at the International Peace Research Institute in Oslo is developing a new and more comprehensive data set, called the Armed Conflict Location and Event Data (ACLED). This innovative data set reports conflict events by date and location, making it possible to use disaggregated subnational data sets. This data set has coded 50 countries to date. However, by keeping with the country/year dyad I can ensure comparability between the different models by both Miguel and other models presented in this chapter.

The rainfall growth data used in my analysis comes largely from two main sources to cover the period from 1981 to 1999. Rainfall data derive

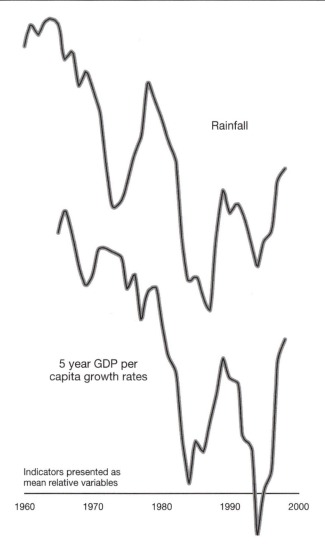

Figure 2.2 Social Vulnerability and Food Insecurity—Rainfall and Economic Growth in Sub-Saharan Africa, 1960–2000. Patterns of economic growth closely follow patterns of precipitation. Since rain-fed agriculture is a large percentage of total agricultural production, changes in rainfall affect economic development. Consequently, when rainfall decreased, so did economic growth. Yet, other factors, such as civil war, low educational attainments and high infant mortality rates, also shaped economic growth in Africa.

Sources: Hugo Ahlenius, "Human Vulnerability and Food Insecurity—Rainfall and Economy in Sub-Saharan Africa," UNEP/GRID-Arendal Maps and Graphics Library, 2008, http://maps.grida.no/go; Barrios Salvador, Luisito Bertinelli, and Eric Strobl, "Dry Times in Africa: Rainfall and Africa's Growth Performance" (Munich: MPRA Paper 5705, University Library of Germany2003, http://ideas.repec.org/p/pra/mprapa/5705.html.

from the Global Precipitation Climatology Project (GPCP) and from the Global Precipitation Climatology Center (GPCC). Both rainfall products consist of a mix of precipitation indicators, combining gauge with satellite data. Both data sets have similar values, and the correlation between the GPCP and GPCC from 1981 to 1999 is significant at the 0.01 percent level. It is interesting to note that measures of water availability, especially water runoff, did not yield significant results. This may be due to the difference in the composition of the runoff measures. Rainfall growth is measured annually, calculating the difference in rainfall from year to year for each country, or:

$$\Delta R_{it} = \frac{R_{it} - R_{it-1}}{|R_{it-1}|}$$

Some of the control variables are the same as in Miguel's model to ensure comparability. They include ethnolinguistic fractionalization based on the Soviet ethnographic index, religious fractionalization based on the CIA Factbook, level of democracy drawn from the Polity IV data set, and the proportion of a country's mountainous terrain.[42] In addition to Miguel, I include a measurement of poverty using infant mortality rates as an indicator for development. The agricultural, economic, and demographic data are drawn from the World Development Indicators database (results of control variables are not shown). Alternatively (and this is the main method in this analysis), I use country fixed effects and country specific time trends to account for additional variation in the models.

Discussion of Empirical Results

At first, I replicated Miguel's first-stage linear-bivariate regression using data covering the years 1961 to 1999 for both samples, the full and the subsample. In both samples, rainfall growth strongly relates to positive economic growth at the 0.01 percent confidence level (see Table 2.1). As expected, the impact of rainfall growth on economic growth is much stronger in the subsample.

In a next step and in line with Miguel, I use rainfall growth as an instrument for economic growth in a regression analysis. As mentioned above, country fixed effects and country specific time trends are included in the model specifications. As already demonstrated by Miguel and colleagues, lagged rainfall growth relates positively to conflict onset. This result is consistent with the 25 and the 1,000 battle-related death thresholds but the effect is larger when using the 25 battle-related death threshold (see Tables 2.2 and 2.3).

The results of the models in Tables 2.4 and 2.5 support the hypothesis that countries with a higher dependency on agriculture have an elevated

Table 2.1 Rainfall and Economic Growth (First-Stage Regression)

Explanatory Variable	Dependent Variable Economic Growth Rate, t			
	Model 1 (Full Sample Based on Miguel et al.)	Model 2 (Full Sample Based on Miguel et al.)	Model 3 (Sub-sample, at Least 30% Agriculture, Value Added [% of GDP])	Model 4 (Sub-sample, at Least 30% Agriculture, Value Added [% of GDP])
Growth in Rainfall, t	0.055***	0.049***	0.090***	0.078**
Country Fixed Effects	no	yes	no	yes
Country-Specific Time Trends	no	yes	no	yes
R^2	0.02	0.13	0.04	0.13
Observations	743	743	352	352

* Significantly different from zero at 90 percent confidence.
** Significantly different from zero at 95 percent confidence.
*** Significantly different from zero at 99 percent confidence.
Source: Edward Miguel, Shanker Satyanath, and Ernest Sergenti. "Economic Shocks and Civil Conflict: An Instrumental Variables Approach." *Journal of Political Economy* 112, no. 4 (2004).

risk of experiencing an armed conflict when experiencing sudden negative changes in rainfall. More specifically, the results (for Model 1) show that the risk of conflict in countries with more than 30 percent of GDP in agriculture is higher by 21 percentage points (see Table 2.4), whereas the risk of experiencing conflict in the full sample is higher by 12 percentage points (see Table 2.2). This dataset includes all conflicts that are above a threshold of 25 battle-related deaths. Using a higher threshold of 1,000 battle-related deaths, the pattern is similar: the risk for the agriculture-dependent countries is higher by 12 percentage points (see Table 2.5), whereas the risk for the full sample is higher by 7 percentage points (see Table 2.3). As a result, in the case of a sharp drop in rainfall, the risk of experiencing violence in a country that is agriculture-based increases by 9 and 5 percentage points, respectively (with Model 2 showing similar results).

In most models, only changes in a lagged rainfall growth variable show significant results. This can be explained by the time lag between the actual precipitation and its subsequent impact on agricultural yields. The only nonlagged variable that is significantly different is the Normalized Difference Vegetation Index ([NDVI]; see Model 7 in Table 2.2, Table 2.3,

Table 2.2 Rainfall and Civil Conflict ($\geq$ 25 Deaths)

Explanatory Variable	Dependent Variable Civil Conflict $\geq$ 25 Deaths						
	Model 1 Using GPCP[1] Data (Miguel et al.)	Model 2 Using GPCP Data (IRI[2])	Model 3 Using GPCC[3] Data	Model 4 Using UNH[4] Runoff Data	Model 5 Using NCEP[5] Data (Miguel et al.)	Model 6 Using FAO[6] Data (Miguel et al.)	Model 7 Using NDVI[7] Data (Miguel et al.)
Growth in Rainfall, t	−0.024	−0.034	0.017	−0.001	0.011	0.002	−0.345***
Growth in Rainfall, t-1	−0.122**	−0.146***	−0.103**	−0.001	−0.055	−0.031*	−0.143
Country Fixed Effects	yes	yes	yes	yes	yes	yes	yes
Country-Specific Time Trends	yes	yes	yes	yes	yes	yes	yes
R^2	0.71	0.71	0.71	0.71	0.71	0.72	0.71
Root Mean Square	0.25	0.25	0.25	0.25	0.25	0.24	0.26
Observations	743	743	743	743	743	607	607

* Significantly different from zero at 90 percent confidence.
** Significantly different from zero at 95 percent confidence.
*** Significantly different from zero at 99 percent confidence.

Source: Edward Miguel, Shanker Satyanath, and Ernest Sergenti. "Economic Shocks and Civil Conflict: An Instrumental Variables Approach." *Journal of Political Economy* 112, no. 4 (2004).

[1] Model 1 uses Global Precipitation Climatology Project data based on Miguel et al.
[2] Model 2 uses Global Precipitation Climatology Project data retrieved from the data library of the International Research Institute for Climate and Society, Columbia University.
[3] Model 3 uses data of the Global Precipitation Climatology Centre.
[4] Model 4 uses runoff data of University of New Hampshire.
[5] Model 5 uses Climate Prediction Center Data based on Miguel et al.
[6] Model 6 uses precipitation data of the Food and Agricultural Organization based on Miguel et al.
[7] Model 7 uses the Normalized Difference Vegetative Index based on Miguel et al.

Table 2.3 Rainfall and Civil Conflict (≥ 1000 Deaths)

	Dependent Variable Civil Conflict ≥ 1000 Deaths						
Explanatory Variable	Model 1 Using GPCP[1] Data (Miguel et al.)	Model 2 Using GPCP Data (IRI[2])	Model 3 Using GPCC[3] Data	Model 4 Using UNH[4] Runoff Data	Model 5 Using NCEP[5] Data (Miguel et al.)	Model 6 Using FAO[6] Data (Miguel et al.)	Model 7 Using NDVI[7] Data (Miguel et al.)
Growth in Rainfall, t	−0.062**	−0.066**	−0.063**	−0.00003	−0.064**	−0.006	−0.310***
Growth in Rainfall, t-1	−0.069**	−0.082***	−0.047	−0.00003	−0.065*	−0.009	−0.180**
Country Fixed Effects	yes	yes	yes	yes	yes	yes	yes
Country-Specific Time Trends	yes	yes	yes	yes	yes	yes	yes
R^2	0.70	0.70	0.70	0.70	0.70	0.80	0.68
Root Mean Square	0.22	0.22	0.22	0.22	0.22	0.17	0.23
Observations	743	743	743	743	743	607	607

* Significantly different from zero at 90 percent confidence.
** Significantly different from zero at 95 percent confidence.
*** Significantly different from zero at 99 percent confidence.

Source: Edward Miguel, Shanker Satyanath, and Ernest Sergenti. "Economic Shocks and Civil Conflict: An Instrumental Variables Approach." *Journal of Political Economy* 112, no. 4 (2004).

[1] Model 1 uses Global Precipitation Climatology Project data based on Miguel et al.

[2] Model 2 uses Global Precipitation Climatology Project data retrieved from the data library of the International Research Institute for Climate and Society, Columbia University.

[3] Model 3 uses data of the Global Precipitation Climatology Centre.

[4] Model 4 uses runoff data of University of New Hampshire.

[5] Model 5 uses Climate Prediction Center Data based on Miguel et al.

[6] Model 6 uses precipitation data of the Food and Agricultural Organization based on Miguel et al.

[7] Model 7 uses the Normalized Difference Vegetative Index based on Miguel et al.

Table 2.4 Rainfall and Civil Conflict (≥ 25 Deaths), Countries with at Least 30 Percent Agriculture, Value Added (% of GDP)

	Dependent Variable Civil Conflict ≥ 25 Deaths						
Explanatory Variable	Model 1 Using GPCP[1] Data (Miguel et al.)	Model 2 Using GPCP Data (IRI[2])	Model 3 Using GPCC[3] Data	Model 4 Using UNH[4] Runoff Data	Model 5 Using NCEP[5] Data (Miguel et al.)	Model 6 Using FAO[6] Data (Miguel et al.)	Model 7 Using NDVI[7] Data (Miguel et al.)
Growth in Rainfall, t	−0.032	−0.021	0.050	−0.003***	0.032	0.040	−0.503*
Growth in Rainfall, t-1	−0.205**	−0.210**	−0.159*	−0.003***	−0.066	−0.162*	−0.250
Country Fixed Effects	yes	yes	yes	yes	yes	yes	yes
Country-Specific Time Trends	yes	yes	yes	yes	yes	yes	yes
R^2	0.68	0.68	0.68	0.68	0.68	0.70	0.68
Root Mean Square	0.29	0.29	0.29	0.29	0.29	0.29	0.29
Observations	352	352	352	352	352	287	295

* Significantly different from zero at 90 percent confidence.
** Significantly different from zero at 95 percent confidence.
*** Significantly different from zero at 99 percent confidence.

Source: Edward Miguel, Shanker Satyanath, and Ernest Sergenti. "Economic Shocks and Civil Conflict: An Instrumental Variables Approach." Journal of Political Economy 112, no. 4 (2004).

[1] Model 1 uses Global Precipitation Climatology Project data based on Miguel et al.
[2] Model 2 uses Global Precipitation Climatology Project data retrieved from the data library of the International Research Institute for Climate and Society, Columbia University.
[3] Model 3 uses data of the Global Precipitation Climatology Centre.
[4] Model 4 uses runoff data of University of New Hampshire.
[5] Model 5 uses Climate Prediction Center Data based on Miguel et al.
[6] Model 6 uses precipitation data of the Food and Agricultural Organization based on Miguel et al.
[7] Model 7 uses the Normalized Difference Vegetative Index based on Miguel et al.

Table 2.5 Rainfall and Civil Conflict (≥ 1000 Deaths), Countries with at Least 30 Percent Agriculture, Value Added (% of GDP)

Explanatory Variable	Dependent Variable Civil Conflict ≥ 1000 Deaths						
	Model 1 Using GPCP[1] Data (Miguel et al.)	Model 2 Using GPCP Data (IRI[2])	Model 3 Using GPCC[3] Data	Model 4 Using UNH[4] Runoff Data	Model 5 Using NCEP[5] Data (Miguel et al.)	Model 6 Using FAO[6] Data (Miguel et al.)	Model 7 Using NDVI[7] Data (Miguel et al.)
Growth in Rainfall, t	−0.115*	−0.105*	−0.109*	0	−0.104***	−0.056	−0.642***
Growth in Rainfall, t-1	−0.124**	−0.118*	-0.086*	0	−0.121***	−0.122	−0.104
Country Fixed Effects	yes	yes	yes	yes	yes	yes	yes
Country-Specific Time Trends	yes	yes	yes	yes	yes	yes	yes
R²	0.67	0.67	0.67	0.67	0.67	0.75	0.66
Root Mean Square	0.25	0.25	0.25	0.25	0.25	0.22	0.25
Observations	352	352	352	352	352	287	295

* Significantly different from zero at 90 percent confidence.
** Significantly different from zero at 95 percent confidence.
*** Significantly different from zero at 99 percent confidence.

Source: Edward Miguel, Shanker Satyanath, and Ernest Sergenti. "Economic Shocks and Civil Conflict: An Instrumental Variables Approach." *Journal of Political Economy* 112, no. 4 (2004).

[1] Model 1 uses Global Precipitation Climatology Project data based on Miguel et al.

[2] Model 2 uses Global Precipitation Climatology Project data retrieved from the data library of the International Research Institute for Climate and Society, Columbia University.

[3] Model 3 uses data of the Global Precipitation Climatology Centre.

[4] Model 4 uses runoff data of University of New Hampshire.

[5] Model 5 uses Climate Prediction Center Data based on Miguel et al.

[6] Model 6 uses precipitation data of the Food and Agricultural Organization based on Miguel et al.

[7] Model 7 uses the Normalized Difference Vegetative Index based on Miguel et al.

Table 2.4, and Table 2.5). This is because the index already contains a time lag, as rainfall does not have an immediate impact on changes in the NDVI.

Yet, the question remains why other rainfall variables, such as the deviation from the mean (the number of years that are below a certain threshold, results not shown), or changes in runoff (an indicator of water scarcity, Model 4) did not yield statistically significant results. Only a sudden sharp drop in rainfall yields statistically significant results. The answer could lie in the capacity of countries to adapt to the adverse consequences of subsequent dry years. Another explanation could be the influx of foreign aid in a given drought period, offsetting the negative economic impact, or simply in the lack of robustness of the model.

These results must be interpreted with caution. The relationship described above may mask other mechanisms of civil war onset. It is hardly the "poor farmer" who goes to war. Other factors, such as ethno-political exclusion and high infant mortality rates, may heighten the risk for armed conflict in countries that have a strong agricultural sector. One explanation relates to the structure of the economy. Countries that lack economic integration and have largely agrarian economies—regardless of oil and diamond deposits—have an elevated conflict risk.

> Natural-resource dependent economies may have weak manufacturing sectors . . . and correspondingly low levels of internal trade. Insofar as internal trade is associated with greater levels of social cohesion and interregional interdependence, the weakness of the manufacturing sector and the fragmentation of an economy into independent enclaves of production may raise conflict risks.[43]

Post-conflict countries such as Sierra Leone or Liberia, for example, have not gone through a process of industrialization, and this has generated clusters of agricultural communities with weak commercial ties. This argument is consistent with the weak-state thesis that low-income, largely agrarian countries have a lower capacity to contain upheaval.[44]

SOMALIA, SUDAN, AND NEPAL: INCREASING INSECURITY?

Statistical analysis is incomplete without supplementing it with case-study analysis. In the following, I briefly refer to Somalia, Sudan, and Nepal, and see whether the general pattern observed in the statistical analysis can be confirmed or rejected. All three countries have a history of conflict and are largely agrarian, although in the case of Sudan, revenues from oil have increased in recent years.

CLIMATE CHANGE AND ARMED CONFLICT
IN SOMALIA

For almost two decades, Somalia has experienced human suffering and warfare. All efforts to form a functioning government have failed. The most recent attempt, the establishment of the Transitional Federal Government, did not succeed in uniting the country and providing peace and security. Armed clashes between Ethiopian-supported government troops and militias of the Islamic Courts Union in Mogadishu have once again destroyed the hope of a prosperous and politically stable Somalia. Because Somalia is largely an arid country, highly susceptible to natural disasters, especially droughts and floods, and because its people have been victims of severe famine in recent decades, it seems evident that annual changes in rainfall have a socioeconomic impact with consequences for the country's political stability and human security.

Somalia's Climatic Challenges

Somalia is part of the East African savannah, which is part of the Sahel stretching from Senegal in the West to Djibouti in the Far East. Its climate is semiarid and hot, reaching 45° Celsius. This environment shaped the historical patterns of livelihoods and the predominantly pastoral life of the Somalis. It is a life where animals and humans live in a mutual, interdependent relationship. One cannot live without the other. The semiarid belt covers the whole of Somalia, although particular riverine areas, such as the Lower Shabelle region or the Juba valley, have fertile irrigated land. The precarious and competitive conditions of the savannah shaped Somali culture and traditions, precarious because of the unpredictability of rainfall and competitive because of scarce resources, such as water and grazing areas.[45]

Somalia is a country that is prone to droughts and erratic rainfall. Most farmers of the riverine and inter-riverine areas depend on rain-fed cropping. In Lower Shabelle—depending on the water flow of the Shabelle River—irrigated and flood irrigated farming constitutes an additional but limited source of agricultural activity. In an environment where rainfall is low and unpredictable, local farmers and pastoralists are pushed to the limits of subsistence when external stresses such as droughts or floods impact them. Additional anthropogenic factors, such as deforestation, depletion of the water table through borehole drilling, and overgrazing can lead to a situation whereby environments become critical.[46] Although this does not support the hypothesis that environmental criticality is a cause of violent conflict per se, it is inevitably intertwined with the political causes of conflict in southern Somalia.

Somalia is a critical region. Rainfall is low, unevenly distributed, and irregular, with mean annual rainfall below 500 mm per year. Rainfall can vary in terms of duration and quantity in areas only a few kilometers apart in a given month, season, or year. Crop failures occur at periodic intervals. As a rule of thumb, one in every five harvests will be a partial failure whereas one in 10 is a complete write-off.[47] Apart from the 1926 to 1929 and the 1973 to 1974 droughts that affected the whole country, droughts occur in specific regions and normally last only for one season. However, about 18 droughts not confined to territory occurred in the past 100 years in Somalia.[48] Rainfall plays such an important role for most Somalis that those affected remembered them with special names. Also, droughts are differentiated according to their severity. *Abaar* ("drought") *neebsooy* ("take a rest") refers to the failure of the *gu* or *deyr* rainy season so that farmers can stay at home and "take a rest" from cultivating activities, while the more severe *abaar nuuhiyi* ("nothing is left") refers to the failure of crops, empty stores, no pasture, and no livestock. Besides, farmers not only fear the failure of rainfall but are also plagued by floods. Though floods are important for irrigation, they can become destructive when exceeding certain levels.

Living in one of the hottest areas of the globe, known as the torrid zone, caused Somalis to develop exceptional skills to cope with this environment. Drysdale argues that pastoralism was the most likely livelihood to evolve in a semiarid region such as the African savannah.[49] Water was crucial for human survival in a hot climate. In the pastoral setting, mobility was important to increase social resilience to cope with the unpredictability of rainfall.[50] Camels can go without water for several days, sometimes weeks, which enabled humans to walk long distances.

Likewise, migration among farmers was equally as important a coping strategy as storing grain. Helander argues that mobility among the Rahanweyn households in the Bay region was essential to survive external shocks.[51] For instance, during the 1998 to 1999 drought, people migrated from the Bay region to Lower Shabelle, where clan connections were strong.[52] If mobility were restricted through either war or legislation, it would increase their social vulnerability, forcing them out of the farming sector into urban centers and wage labor. The fragility of the relationship between the Somalis and their environment, together with population growth, led to the assumption that the Somali conflict relates to changes in rainfall patterns. However, the Somali case is more complicated than reducing the conflict to explanations of resource scarcity and population pressures.

Explaining Somalia's Trajectory

Apart from climatic changes and variability, several other factors influence warfare in Somalia. To list just a few, the role of political leadership, the fragmentation of the Somali clan system, external interests, the unjust distribution of economic resources such as charcoal, and the colonial legacy are all factors that shape political violence in Somalia.[53] Some argue that ethnicity, so commonly invoked as an explanation of conflict in contemporary African states, seemed less relevant in the Somali context. Yet, there are perceived differences based on clan lineage that Somali warlords used to lobby for political and military support. This instrumental use of ethnicity has divided the Somali society and weakened a former "pastoral democracy."[54]

In the absence of an internationally recognized government, which could generate wealth from outside, Somali elites started focusing on domestic resources. Environmental change and drought will certainly reduce the available resources, thereby acting as an additional stress factor. It became clear in Somalia that those who benefit from the short-term profits of resource extraction, such as charcoal production, fisheries, large-scale irrigation of cash crops, diversion of foreign aid, and controlling strategic assets such as air- and seaports, benefited from Somalia's lawlessness. Conflict largely arose over taxation of either export crops, such as bananas, or local resources, such as land and water. Here, the war has destroyed the plenty of the region, exacerbating existing tensions between small-scale and large-scale farmers over the distribution of scarce resources, such as water. In addition, traditional institutions of resource sharing among farmers and pastoralists became dysfunctional. This reduced the ability of local communities in coping with external shocks, such as droughts, floods, or market failures.

In Somalia, economic interests have become significant in the perpetuation of the civil war; some authors underline this point.[55] Small but influential groups thus come to have an economic interest in prolonged conflict because they benefit from lawlessness. This viewpoint affirms that it can be misleading to associate war with complete collapse or breakdown of an economy—although it may certainly skew the development of an economy. The Somali pirates are good examples of Somalia's power vacuum. No authority exists that could patrol the Somali coasts to protect the country's rich fisheries while preventing pirates from entering and capturing foreign vessels. The Somali case is complex and multifaceted; what is needed is a stronger political commitment from the African Union, the neighboring states, and other international actors, to build a functioning government that has legitimacy and authority over its territory.

Moreover, future droughts will not necessarily affect Somalia's population evenly. Alexander de Waal argued that the famine from 1991 to 1992 in southern Somalia was highly selective: The people most affected by the famine were farming communities and internally displaced persons.[56] Once again, this is a good example of how climate change impacts will affect people depending on their vulnerability. In sum, climate change impacts are only one of the many factors that shape conflict and human insecurity, making it more difficult to build a prosperous and peaceful future.

CLIMATE CHANGE AND ARMED CONFLICT IN SUDAN

Sudan, as the largest country in Africa, has a long-standing history of armed conflict. When the Comprehensive Peace Agreement (CPA) was signed in 2005, there was hope for a prosperous and politically stable Sudan. Instead, massive killings in the Darfur region triggered a humanitarian catastrophe. A recent study published by the UN Environment Programme claims that "there is a very strong link between land degradation, desertification and conflict in Darfur."[57] Sudan, a highly agrarian country, has long been dependent on rainfall, which corresponds to the vicissitudes in the country's economic growth pattern, especially until the late 1990s when oil gained importance (see Figure 2.3). The country had an exceptionally good year in 1994, with a bumper crop and no reports of major armed conflicts in either 1993 or 1994. This enabled Sudan's economy to prosper in relative stability. However, little attention has been paid to the economic circumstances of this region that largely depends on erratic rainfall, making it vulnerable to the adverse consequences of climate change.

Conflict in Sudan

Sudan gained its independence from Britain in 1956, but since then has spent little time free from conflict. Weak institutional structure during

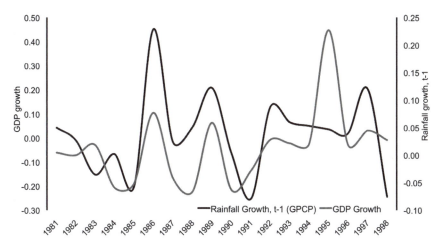

Figure 2.3 Rainfall Growth and Economic Growth in Sudan, 1981–1998. Economic growth trails rainfall patterns in the Sudan, a country that has a large agricultural sector; 1994 was a particularly good year in terms of agriculture, improving the financial situation in the following year.

Sources: Global Precipitation Climatology Project (GPCP), International Global Energy and Water Cycle Experiment (GEWEX) Project Office (IGPO), Silver Spring, MD; and World Development Indicators, World Bank, Washington, DC.

colonial times did not improve after independence. Divides within the country grew and ongoing civil war (1962 to 1972, 1983 to 2005, and more recently in Darfur) did not allow time to invest enough resources to address the country's needs. Differences between the North and South, partly created during colonial rule, are also largely due to regional environmental variation and preexisting cultural and religious groups in each region.

The North has been populated generally with Arabic Muslims and has always been more developed. This is in direct contrast with the very underdeveloped southern Sudan, which has been inhabited largely by Africans who are either animist or Christian. Observers of the recent conflict in Darfur also phrased it in the terms of ethnicity: black African farmers fighting Muslim herders, which, of course, is an oversimplification of the complex societal dynamics and power-relations.

Sudan's Vulnerability to Climatic Variability

Sudan's economy, similar to most African nations, relies heavily on its agricultural sectors. Agriculture represents roughly 80 percent of the work force and has for decades. In 1996, agriculture accounted for 48 percent of the country's GDP. In 2005, this number decreased to 39 percent; the contribution of crude oil exports has steadily increased since Sudan began exporting this natural resource in 1999. Though the contribution to GDP has decreased by roughly 10 percent, the percentage of the population employed by agriculture has remained at 80 percent. Sudan also boasts the second largest animal population in Africa, which annually contributes notably to the nation's GDP. Civil war has marred Sudan since 1983. Despite this, the country's wealth of natural resources (oil) has allowed for recent GDP growth of close to 7 percent.

Even with recent economic success from crude oil exports, Sudan's population has not shifted from being overwhelmingly agrarian, and remains largely dependent on rain-fed farms. Sudan faces several environmental issues. All but one has to do with water. Inadequate supplies of potable water, soil erosion, desertification, and periodic drought all plague Africa's largest country and fifth largest population. An estimated southward shift of 50 to 200 km in the boundary between semidesert and desert has occurred since scientists first collected records in the 1930s.[58] Whereas an estimated 42 percent of the total area is considered cultivatable land, only 7 percent of that portion is actually cultivated. Average rainfall in Sudan varies greatly by region, from annual rainfall of less than 25 mm in the North to nearly 1500 mm in the South. Figure 2.4 shows that Darfur has experienced below-normal annual rainfall for several years, increasing tensions over scarce water resources and pasture. Vast swamp regions in Sudan also affect the water stability in the country, greatly increasing the amount of evaporation that occurs annually (roughly three times as much water

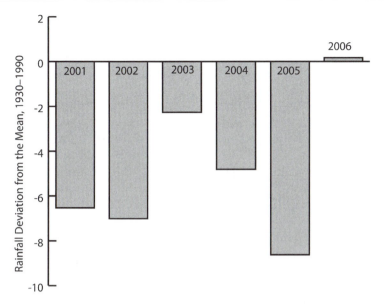

Figure 2.4 Rainfall Deviations in Darfur, 2001–2006. Darfur experienced intense civil unrest and war starting in 2003 and lasting until 2010, when a cease-fire was signed. The years before and following the conflict show low levels of rainfall, a critical resource required for economic development and human security.
Source: The International Research Institute for Climate and Society/Lamont-Doherty Earth Observatory (IRI/LDEO) Climate Data Library, Dataset: Global Precipitation Climatology Centre, Landsurface Monitoring Product 1.0 Degree.

evaporates due to the swamps as is annually available for use).[59] Annual rainfall has been declining for some time, and publications link this to weak economic performance. In addition, the war left large areas contaminated with unexploded ordnance and targeted natural resource destruction, such as deliberately felled trees.[60] Further, more than 5 million internally displaced persons contributed to environmental damage and resource scarcity.

The Sudan Case: A Complex Challenge

More recently in Sudan, natural resource wealth—mainly from oil—offers large resource rents to elites by providing the financing needed to start or to sustain conflict. These rents provide strong incentives for "peace spoilers."[61] As mentioned earlier, the dependence on natural resource production weakens the state structures that redistribute wealth, and are less reliable and competent to provide public goods. Weak states, in turn, are at higher risk of civil war because they rely on natural resources for revenue rather than on taxation, have weaker state structures, and are thus less able to contain violence. In addition, they tend to be less democratic; they need not be accountable to the public because natural resources provide large rents that preclude the need for taxation.[62] Clearly, the causes that trigger, sustain, and prolong recurring conflict are multidimensional and cannot be reduced to a single variable.

Most nations struggle to deal with just one major problem. Sudan has been dealing with many. Along with years of violent conflict, lack of national infrastructure, corrupt governance, population pressures, and a high disease burden, rainfall levels have continued to decline. The latter has only exacerbated the problems associated with the former. Moreover, those problems initially listed have left Sudan unable to rectify the fact that it is an agrarian nation, reliant on precipitation.

CLIMATE CHANGE AND HUMAN INSECURITY IN NEPAL

Nepal is a country that is largely dependent on climate-sensitive sectors, such as rain-fed agriculture; its ecosystem is fragile and the dramatic topography makes the country prone to flooding.[63] The country's institutions are weakened and fragile after years of civil turmoil with remaining pockets of insecurity, thus undermining efforts to adapt to climate change. Adaptation to climate change is costly, and Nepal desperately needs financial resources to rebuild the country, reintegrate ex-combatants, and pay salaries to civil servants.

Statistical studies show that the years prior to a high-intensity conflict in 2002 were drier than normal.[64] The occurrence of a significant drought in the year prior to the conflict outbreak weakened the local economy by creating more unemployment and desperation among farmers. This was likely to be one of the factors that lowered the opportunity costs of potential rebels to join the Maoist movement. There were other factors at work, such as intergroup inequalities in the Western part of the country. Research on spatial (i.e., geographical) horizontal, intergroup inequality in Nepal finds that horizontal inequalities robustly explain the intensity of the Maoist insurgency (although not the incidence), with a statistical association between district-level relative deprivation in terms of human development and the incidence of violent deaths in that district.[65]

Significant political change marked recent years: In a historic moment in April 2006, Nepal abolished its monarchy. The Maoist insurgency that started in 1996 ended after a nine-year civil war that left thousands of people dead. Following extensive negotiations, the government and the Maoists signed a Comprehensive Peace Accord in November 2006.[66] A federal republic was established, creating a constitutional assembly. The interim parliament, with a Maoist majority, elected a president in July 2008. This peace is fragile, threatened by several armed political groups that oppose the Peace Accord. What is more, climate change is likely to affect the agrarian-based economy negatively, and will influence the tourism industry when glaciers retreat or cause devastating glacier lake outbursts and subsequent flash floods. Although Nepal's per capita greenhouse gas emissions are negligible, the country is most affected by the negative consequences of climate change, such as flash floods, glacier lake outbursts, and rising average maximum temperatures with a subsequent retreat of glaciers, and least able to adapt to a changing environment.

Increase of Flash Floods

Flash floods affect thousands of people every year in the Himalayan region. Intense rainfall is the main cause of flash floods, although other types of floods such as those caused by rapid snowmelt also occur. As climate models project an increase in monsoon precipitation, an increase in flash floods is likely. Similarly, the increase in intensity of precipitation is also a possible trigger for flash floods.[67] In turn, flash floods destroy crops, thereby inducing resource scarcity in the affected regions.

Glacier Lake Outbursts

Much attention has focused on glacier lake outbursts. As glaciers retreat, they release water that remains trapped behind now-exposed moraines. Moraines that act as dams can break, resulting in a glacier lake outburst and subsequent flooding with catastrophic consequences for lower-lying populations. Although relatively rare events, in 1984, a glacier lake outburst in the Langmoche valley in the Khumbu region washed away bridges, agricultural land, homes, and people; it was felt more than 90 km downstream. It also destroyed a nearly completed hydropower project and caused approximately US$1million damage.[68]

Temperature Changes in the Higher Altitudes

Experts and local communities have already observed changes in temperature averages. According to the International Centre for Integrated Mountain Development (ICIMOD), warming in Nepal is on average at 0.6° centigrade per decade, higher than the global average.[69] Most importantly, the warming in the Himalayas was far greater at higher elevations. As one expert notes: "The mid-hills area used to be the main habitat of people and the impact of climate change is not very visible" but "those closer to the higher mountains seem to have greater awareness in terms of less snow coverage. The Sherpas are worried that there is no snow in the winter, so will [they] get water?"[70]

Changes in Runoff of Major Himalayan Streams

Changes in the length and intensity of precipitation and temperatures not only influence snow cover but also affect downstream populations. According to ICIMOD, about 1.3 billion people live in river basins originating in the Himalayan region. As stated in Table 2.6, about 178 million people live in the Indus river basin, which is likely to experience up to a 45 percent river flow change in snow and glacier melt. With one of the highest population densities in the region, the Indus basin is populated largely with poor communities. A decrease in river discharge coupled with population growth and poverty will expose more people who are mostly agrarian to water scarcity. It is in this context that changes in the Himalayas are not isolated events: "Everybody looks at the glacial lakes and melting glaciers, but they are the tip of the iceberg, representing a larger, dramatic change with consequences downstream."[71]

The Future Remains Uncertain

Nepal's challenges ahead are clear. The main immediate risks resulting from climate change are flash floods, glacier lake outbursts, changes in crop yields and type due to changing temperatures, and in the long term, we can expect significant changes in downstream water flows, affecting millions of people downstream. The country, which has a history of conflict, is now in a fragile and dangerous transition to a more peaceful and stable society. Although Nepal's per capita emissions are low, the country is engaged in contributing toward a low-carbon society. One example is Nepal's successful community forest-management program, which has led to additional carbon sequestration and watershed protection and created a buffer zone between protected areas and communal agricultural land. The described effects of climate change will put more pressures on an already fragile economy, elevating the risk of conflict recurrence. Adaptation to climate change requires desperately needed resources to rebuild the country and to reintegrate the thousands of ex-combatants.

CONCLUSION

Certainly, all three countries, Somalia, Sudan, and Nepal, are agrarian-based, which makes them susceptible to sudden drops in rainfall and subsequent droughts. Countries with a higher dependency on the primary sector are likely to lose economic income due to drought. In turn, low incomes elevate the risk of conflict. However, given the methodological pitfalls of the replication of Miguel's study, and the omission of important intervening variables, there is weak evidence to support a direct relationship between drought and civil war. Among others, ethno-political exclusion seems to be one of the more robust predictors of civil war onset.[72]

Table 2.6 Main Rivers of the Himalayan Region

	River		River Basin			
	Mean Discharge (m3/s)	Glacial Melt in River Flow (%)	Area (km^2)	Population x1000	Population Density	Water Availability (m^3/person/year)
Indus	5,533	44.80	1,081,718	178,483	165	978
Ganges	18,691	9.10	1,016,124	407,466	401	1,447
Brahmaputra	19,824	12.30	651,335	118,543	182	5,274
Irrawaddy	13,565	Small	413,710	32,683	79	13,089
Salween	1,494	8.80	271,914	5,982	22	7,876
Mekong	11,048	6.60	805,604	57,198	71	6,091
Yangtze	34,000	18.50	1,722,193	368,549	214	2,909
Yellow	1,365	1.30	944,970	147,415	156	292
Tarim	146	40.20	1,152,448	8,067	7	571
Total				1,324,386		

This shows their mean discharge volume and share of glacial melt. The percentage of glacial melt in the Indus River, for instance, on which some 178 million people depend for water, is up to 45 percent. With climate change and a subsequent retreat of glaciers, we can expect a change in river flow affecting millions of people downstream.

Source: Adapted from IUCN/IWMI, Ramsar Convention and WRI, 2003; Mi and Xie 2002; Chalise and Khanal 2001; Merz 2004.*

*The hydrological data may differ depending on the location of the gauging stations. The contribution of glacial melt is based on limited data and should be taken as indicative only.

Accordingly, predicting an increase of warfare in Africa and Asia based on declining rainfall as suggested by the climate change scenarios of the Intergovernmental Panel on Climate Change is problematic. First, the variation of the predictive models is very large, making it difficult to use the models to forecast where conflicts will likely take place. Second, important socioeconomic control variables, such as GDP per capita or population density, are also subject to change, and we do not know the direction of this change. Third, the causes of conflict are complex and multifaceted. Analyzing the spatial correlation of rainfall anomalies and conflict hotspots in Africa may be a useful policy instrument to understand where droughts exacerbate existing tensions over renewable resources. Accordingly, drought insurance schemes as introduced in Malawi may be the way forward to deal with the adverse political consequences of increased climatic variability.

It remains an open question whether human-induced climate change manifested in droughts will lead to more conflict. Research by scholars at Columbia University's International Research Institute for Climate and Society has demonstrated that the twentieth-century drying of the Sahel, where many of the conflict zones are located, can be attributed to anthropogenic activities (see Figure 2.2). Although it is beyond the scope of this chapter, it is interesting to follow the question of to what extent anthropogenic climate change will influence the risk of civil conflict in the future. As a possible reflection of this uncertainty, the IPCC *Fourth Assessment Report* makes little reference to conflict or political instability.[73] Given the fact that some countries in particular in the developing world will continue to depend on agriculture, global climate change may, however, affect people in more serious ways than we have anticipated.

CHAPTER THREE

Natural Disasters and Security Implications

The least movement is of importance to all nature. The entire ocean is affected by a pebble.

Blaise Pascal (1623–1662), French mathematician
and philosopher

INTRODUCTION

Scientists agree that anthropogenic climate change is already a reality. The Earth has become 0.74° Celsius warmer over the last 100 years.[1] There is already the perception that the intensity and potentially the frequency of natural disasters is increasing, especially droughts and episodes of high temperatures, and some groups suggest that this has lead to an increase in tropical cyclones, though there is no definite proof of that to date. More storms and droughts in a warmer climate will affect more people with negative consequences for human security. When people leave their homes, forced to migrate elsewhere, they become (environmental) refugees, with social and political repercussions for other regions and countries. This chapter largely focuses on the implications for the people-centered concept of human security, meaning the protection from any type of threat. This is distinct from the rather narrow concept of security used by realism, a tradition in thinking in international politics that is concerned with war and the use of force where the main actors are states. In this context, security is compromised by threats to sovereignty.

There are two types of natural disasters: They are either geological or hydro-meteorological in nature. Climate change is most likely to impact hydro-meteorological disasters. One sign of a changing climate is the fact that the number of hydro-meteorological disasters, such as tropical cyclones (with associated flooding) and droughts has increased over the past decades, according to the Centre for Research on the Epidemiology of Disasters (CRED).[2] Overall, floods are the most frequent type of disasters, followed by droughts.[3] Global weather patterns and changes in the concentration of CO_2 in the atmosphere influence both types. The year 2007 alone recorded 414 disasters, killing 17,000 people.[4] The majority, 90 percent, of these disasters occurred in Asia.[5] There, a large proportion of the population lives close to coasts and rivers. This enables Asia's emerging economies to access international markets, as well as making them vulnerable to coastal flooding and tropical cyclones. Scientists estimate that a staggering 82 percent of the world population lives in flood-prone regions.[6] In particular, the poor, who are unable to move to safer places, will become more vulnerable. Adaptation, such as coastal sea defenses, will be more difficult for poor countries, such as Vietnam or the Philippines, whereas developed countries have the economic capacity to prepare, respond, and adapt to natural disasters. For instance, Hurricane Katrina in 2005 caused great economic damage and claimed more than 1,800 lives but it did not affect the country's economic performance. Perhaps a sign of successful worldwide adaptation is the fact that the number of disaster deaths has fluctuated over the past decades, despite the fact that the number of hydro-meteorological disasters steadily increased.[7]

There are other security implications in addition to impacts on human health or human well-being: The adverse effects of climate change, such as flooding, storms, and droughts, add pressures to societies in already weak and fragile states. As mentioned earlier, a statement of a recent policy report endorsed by 11 retired U.S. generals and admirals argues that climate change can act as a "threat multiplier."[8] Although rich countries are largely responsible for anthropogenic climate change, natural changes have less impact there, though the case of Japan illustrated in this chapter shows the potential loss of economic productivity due to future storm increases. In industrialized countries, climate change mitigation has another dimension: Most developed countries want to protect their national interests, especially in the area of energy security. For instance, the United States seeks predictable and stable regimes in Africa because of its strategic interest in natural resources. In 2015, Africa may provide 25 to 40 percent of the country's oil and be a supplier of crucial minerals such as chromite, bauxite, platinum, and manganese.[9]

One of the likely effects of global warming and the subsequent warming of the oceans is that they will generate an increase in the intensity and in some regions the frequency of tropical cyclones.[10] A 30-year satellite record of tropical cyclones confirms this trend.[11] However, the accuracy of satellite-based pattern recognition remains a matter of intense debate.[12] What we know is that sea surface temperatures increased by 0.5°C between 1970 and 2004.[13] Though we should be cautious using global averages, the magnitude of temperature variability is by far greater than trends in mean temperatures. Some areas of the oceans are cooling while others are warming.[14]

Another adverse impact of climate change is the predicted increase in droughts. A policy report claims that in general, wet places are likely to become wetter, dry regions drier.[15] This may be a gross generalization but we have evidence that the Sahel drought in the 1970s and 1980s has been associated with human activities. In countries where the economy is climate-sensitive and a large proportion of the population works in agriculture, the risk of losing economic income due to drought is high. This largely applies to the African continent (see Figure 6.1). Drought not only affects crops and livestock, it also kills men, women, and children. Furthermore, mortality causes covariate economic losses, such as the loss of household assets, lost income, and lost productivity.[16]

This chapter depicts in detail two major human and economic security issues, the impact of tropical cyclones with a regional focus on Asia and the Caribbean, and the impact of droughts and its security implications, largely in Africa. I divide this chapter into three parts. The first section discusses the concepts of social resilience and vulnerability, and explains why these terms are relevant for climate change and human security interactions. The second part deals with the impact of tropical cyclones on economic and human security and to what extent climate change increases vulnerability and reduces resilience to tropical cyclones. I highlight regions that are likely to be most affected by an increase in frequency and intensity of natural disasters, largely tropical cyclones, as a result of a warming of the planet. More specifically, this follows evidence and illustrations from Japan, Haiti, and the Dominican Republic. The last part—though in less detail as some issues of this topic have been already covered in the previous chapter on resource scarcity—is concerned with the effects of droughts on human well-being, with special reference to Somalia.

KEY CONCEPTS: SOCIAL RESILIENCE AND VULNERABILITY

The notion of social resilience and vulnerability are central to the question as to what extent societies are able to adapt to more intense natural disasters. Literature on adaptation barely acknowledges the fact that people in poor countries historically developed sophisticated mechanisms to cope with environmental stress. In many African countries, poor governance and civil unrest disrupted these mechanisms, making the population vulnerable to natural disasters.

Social Resilience

In environmental sciences, the term *social resilience* is used in the context of risk assessment, and is distinct from the meaning in ecology.[17] Social resilience is used to describe the extent to which a community or group of people copes with and adapts to external stress and environmental change. External stress means forced adaptation to the changing physical environment and the disturbance of individuals' or groups' livelihoods. Accordingly, the impact of natural disasters and environmental scarcity on human well-being depends on social resilience. In other words, social resilience determines the level of social vulnerability. In this context, does social resilience reduce the risk of adverse impacts of extreme weather events? Does social resilience strengthen the society's capacity to distribute risk equally at the receiving end? Social and ecological resilience are intertwined when people and their economic activities depend on the natural resources of an ecosystem. If the ecosystem were not resilient to external stress or environmental change, this would affect the depending communities and their economic well-being.

Adger and O'Riordan identify three main proxies for social resilience: Economic indicators, institutions, and demographic factors.[18] First, economic growth, income distribution, and variability of income sources influence social resilience. Economies that are dependent on single resources, such as minerals or natural products, are affected directly by changes in world market prices and cannot buffer financial losses through diversification of market products. If the economy relies on renewable resources, extreme weather events, such as floods or droughts, can increase the risk of high economic loss and consequently decrease economic resilience. Second, the resiliency of institutions depends on different factors. Trust in the legitimacy of institutions plays a major role. If people are forced to adapt to a changing environment that jeopardizes their livelihoods, institutions gain credibility by serving the people's needs to cope with change. This requires political emphasis on environmental risk management, response measures, and adaptation strategies. Third, demographic factors indicate the level of social resilience. Migration can be a sign of stability and resilience; but it depends on the type of population movements.[19] Political instability and economic factors such as high unemployment rates can cause migration; it does not necessarily correlate with environmental change. Even conservation projects that aim to increase ecological resilience through conserving biodiversity can lead to displacement of indigenous people.[20] Defining "push" and "pull" factors can give an indication of the motives of migrants. According to Adger and O'Riordan, migration can reduce the risk of resource dependency at the household level and therefore increase resilience. On the other hand, external stress and migration can indicate the breakdown of social resilience.[21]

Social Vulnerability

The concept of social vulnerability is essential in the context of global environmental change, since it determines how and to what extent people are able to cope with external shocks. The assessment of vulnerability became important in defining the scale of risks, such as natural hazards and manmade disasters. The factor of vulnerability indicates why people in specific geographical regions are more or less able to cope with these risks. A people-centered or anthropological approach is needed to identify the underlying social factors that increase social vulnerability. The term *social vulnerability* emphasizes the importance of focusing on not only measurable natural indicators, such as droughts, coastlines, and geomorphic and geographical factors, but also takes social, political, and economic indicators into account, such as poverty, property rights, entitlements, and access to decision-making processes.

To gain a better understanding of the term *social vulnerability*, it is useful to look at different definitions offered in the literature. It is worth considering the Latin word *vulnerabilis*, which was used to describe a soldier lying wounded on the battlefield, and therefore at a high risk for further attacks.[22] This meaning implies that a vulnerable group or region is already weak and is therefore sensitive to external stress. Three main elements are essential for the term *vulnerability*: The first element contains the present-day state of the individual or the group, the second is the extent of external stress, and the third element refers to the ability to cope with and adapt to external stress. Along these lines, Blaikie and his colleagues suggest considering social, political, and economic factors, shifting the emphasis from purely physical factors as determinants of vulnerability.[23] Some people have greater access to capital or the decision-making process in government institutions to mitigate the negative impacts of disasters. Social factors such as poverty can increase vulnerability but are not a prerequisite.[24]

In a similar way, economic and social entitlements, endowments, and access to resources must carry weight in the assessment of social vulnerability. In this respect, it is revealing to consider people's ability to gain and control access to valuable resources. For instance, in many largely subsistence-oriented societies in sub-Saharan Africa, social support networks are crucial for the survival of the individual. A society that depends mainly on natural resources in a subsistence economy binds together by mutual interest based on reciprocity. For instance, in Somalia the notion of reciprocity is an important component of the community, as people confront the harsh living conditions of a semiarid country. In Somalia, these redistribution mechanisms play an important part within the clan structure. Any member of the same clan can claim support—either moral or material—from his or her kinship group. This form of "moral economy," based on mutual trust, egalitarian principles, and tradition, is inherent to Somali society. Political scientists and anthropologists have relabeled these networks of reciprocity and social support as "social capital."

According to James Coleman, social capital consists of three components: obligations and expectations, information channels, and social norms.[25] It is apparent that the concept of *social capital* combines concepts deriving from sociology and economics. Sociologists would interpret society as a network of norms, values, expectations, and obligations and their interplay. On the contrary, economists interpret society as consisting of individuals who act according to the rational actor model in which humans maximize utility for self-interest. Coleman links the second component of his definition of social capital, information, to the economic concept of transaction costs influencing the level of social capital. However, the borrowing of certain elements from two distinct disciplines also has its pitfalls. Ben Fine criticizes the concept of social capital, since it neglects negative aspects of society.[26] Is it possible to explain the phenomenon of criminal behavior with the concept of social capital in a holistic fashion? Criminal potential, social disorders, or disabilities should be included in the account that constitutes social capital.

With reference to natural disasters, social capital becomes relevant in coping with the floods and droughts. The emphasis is on social networks that provide access to a group's resources. Social capital becomes, therefore, a means of social connections to access resources. In this respect, the analysis of a group's social capital can help to illuminate its vulnerability to external shocks such as natural disasters.

MORE TROPICAL CYCLONES: MORE INSECURITY?

One hydro-meteorological phenomenon that is responsible for floods and great socioeconomic losses are tropical cyclones. Tropical cyclones are also referred to as hurricanes when they occur in the Atlantic Ocean or typhoons when they occur in the Pacific Ocean. Hurricanes, typhoons, and cyclones are synonyms for the same phenomenon, the tropical cyclone. These storms are typically associated with heavy winds, storm surge, and flooding. They develop in the tropics when water temperatures reach more than 26°C. This critical sea surface temperature threshold is widely accepted as a requirement for the formation of cyclones.[27] But the function of higher sea surface temperatures and an increase in the intensity of cyclones does not follow a linear relationship. Other factors influence the storm's size and intensity.[28] These are, among others, vertical shear and mid-tropospheric moisture.[29] Though there is no global trend in an overall increase in the numbers of tropical cyclones (besides the North Atlantic, which shows a significant increase), there was an increase in the strongest category (4 and 5) of tropical cyclones in the past. The number has almost doubled from the 1970s in all ocean basins, though they are also believed to follow cycles and currently it is not clear how these are influencing the overall situation.[30]

The environmental impacts of tropical cyclones can affect a country's security and economic stability in several ways. In most countries, strong winds and associated flooding disrupt human activities and cause severe economic damage by destroying property, infrastructure, and port facilities. In low-income countries, such as Bangladesh, the Philippines, or Haiti, tropical cyclones can claim thousands of lives. There are also health implications. Waterborne diseases and new breeding sites for disease vectors such as mosquitoes often claim several thousand lives in the aftermath of the disaster event. For example, in Honduras in 1998, an additional 30,000 cases of malaria and 1,000 cases of dengue fever were reported following Hurricane Mitch.[31] In Africa today, 450 million people are exposed to malaria with 1 million dying each year. Though a drier climate can reduce this number, an increase in tropical storms is likely to increase the number of people affected by diseases such as malaria or cholera.

The fact is that tropical cyclones claim thousands of lives and cause enormous economic damage. The deadliest tropical cyclone in human history, cyclone Bhola in 1970, devastated the provinces of East Pakistan (today's Bangladesh), claiming between 300,000 and 500,000 lives.[32] This triggered a series of events with serious political consequences. The late and inadequate response of the Pakistani government supported the formation of a rebellion and the establishment of a new country, Bangladesh, in 1971. More recently, in 2005 in the United States, Hurricane Katrina caused major damage estimated at US$30 billion[33] and left more than 1,800 people dead, triggering a debate about whether such extreme weather events will occur more frequently in the future. In the Pacific, too, typhoons have been responsible for large economic losses and claimed hundreds of lives. In 2006, typhoon Durian left 800 people dead in the Philippines alone.[34] Even in countries such as Japan, where loss of life due to tropical cyclones is rare, the economic damage has been large.

Tropical cyclones largely affect populations who live in low-lying coastal areas. Countries in Asia, mainly China, the Philippines, Vietnam, and Indonesia, are most likely to be affected because they have large populations living in low-lying coastal areas and a high vulnerability to tropical cyclones. What is more, the combination of an increase in stronger tropical cyclones and sea-level rise will lead to more flooding of low-lying river deltas in Asia, Africa, and Europe.

Assessing the Damage of Tropical Cyclones

By measuring the meteorological frequency and intensity of tropical cyclones in the past, we can predict future trends. Although scientists generally agree that tropical cyclones are likely to increase in intensity, there is yet no consensus on the future frequency of these events.[35] However, it is possible to identify trends in the costs associated with cyclones, and to see whether these costs are on the rise. The economic costs of natural disasters are indeed enormous and increasing: The annual average global costs

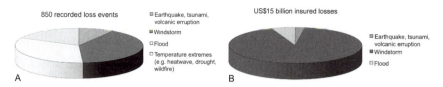

Figure 3.1 Insured and Total Losses in 2006. The majority of all 850 recorded loss events in 2006, 40 percent, were caused by windstorms (A). Windstorms are also responsible for the greatest damage, with 79 percent of the total US$15 billion insured losses in 2006 (B).

Source: Adapted from Munich Re Group. "Topics Geo: Natural Catastrophes 2006" (Munich: Munich Re Group, 2007).

of weather-related natural hazards have increased from US$8.9 billion from 1977 to 1986 to US$45.1 billion from 1997 to 2006.[36]

Reinsurance companies like Munich Re rate cyclones and associated flooding as the most costly natural disaster today. In 2006, windstorms were responsible for 91 percent of total losses from natural disasters, as demonstrated in Figure 3.1. Only 9 percent of losses were due to volcanic eruptions and earthquakes.[37] Moreover, in 2006, windstorms including tropical cyclones accounted for 79 percent of total insured losses, an equivalent of US$13 billion. In the case of Japan, typhoon Shanshan caused losses of US$1.2 billion in 2006.[38] For 2006, an estimated 40 percent of all recorded loss events were associated with windstorms.[39]

Measuring economic damage due to cyclones is a complex problem. Nordhaus lists three reasons for this: First, the impact of maximum wind speeds on damage is nonlinear. Physical damage is lower for low wind speeds and increases sharply with maximum winds. Second, not all tropical cyclones last the same length of time, since cyclone lifetime increases with maximum wind speeds. Third, tropical cyclones above a certain threshold are rare events.[40] Moreover, physical damage is more likely at the point of nonlinear failure. Given these complexities, it is rather difficult to predict and measure the real damage of tropical cyclones.

Evidence from Japan

Even in countries with sophisticated early-warning systems and disaster proven infrastructure, tropical cyclones lead to disruptions of daily life and cause great economic damage. In this context, an increase of tropical cyclone intensity impacts on human security. Countries like Japan, Taiwan, and China, which are affected by tropical cyclones in the Pacific, have experienced great physical damage and other, indirect economic consequences.[41] These are the loss in economic productivity caused by the downtime in the public transport system or other important infrastructure, such as sea ports. A caveat of the existing literature is its focus on the physical or first-order damage of tropical cyclones and associated storm surge damage without calculating the nondirect effects of these storms, such as the disruption to infrastructure or health implications due to flooding. The following calculates the nondirect economic effects for Japan in productive man-hours lost when a certain area is hit by a tropical cyclone. To calculate the economic loss due to a future increase in typhoons, it is important to understand where economic activity is located.

Projecting Climate-Induced Indirect Economic Loss in Japan

Tropical cyclones are events with geographical boundaries. They occur in space and time. Often, they do not pass over inhabited land. But when they do, they are likely to cause disruptions, or worse, they can claim lives. To measure the economic impact of a future increase in typhoons, we need to know where the economic activity is located.[42] In Japan, the highest concentration of economic activity is concentrated in the coastal regions of the East Coast of Central Japan. Across most countries, distribution of the value of economic output is uneven. Urban centers are often the engine of economic growth. The damage caused by tropical cyclones depends on several factors, such as the location of economic activity, number and intensity of tropical cyclones, the region's topography of the cyclone passage, in addition to other geographical characteristics such as land-use patterns.

There are other socioeconomic factors that shape a country's exposure risk to tropical cyclones, such as population growth or economic growth. Economic growth reduces vulnerability to natural disasters, while growth in population, especially in developing countries, will increase vulnerability by exposing more people to more stress from natural hazards, such as the destruction of infrastructure or coastal flooding. In addition, tropical cyclone damage and the cost to adapt to climate change may lower overall gross domestic product (GDP) growth.[43] Another factor that shapes a society's vulnerability is its capacity to adapt. Climate change is a gradual process and will not happen overnight. Hence, climatic changes will probably be gradual, and this will enable societies to adapt to the situation.[44]

A group of researchers, including Miguel Esteban and Tomoya Shibayama of Waseda University, and the author, estimated that the annual economic damage for the Japanese economy due to productivity loss from typhoons with a sustained wind speed of at least 30 knots will account for 0.15 percent of the Japanese GDP in the year 2085 (based on 1990 GDP figures).[45] To illustrate better, this equals approximately US$60 per capita.[46] In absolute figures, this would amount to more than US$7 billion (1995 purchasing power parity [PPP]). The assumptions of the calculations are that in areas that are affected by typhoons with an area of sustained wind speeds higher than 30 knots, most human economic activities come to a standstill, and that typhoon tracks and socioeconomic variables do not change in the future. Such a simplistic study perhaps does not reflect reality as much as studies accounting for socioeconomic future changes, but it is easier to interpret.

The study shows that future loss potential is particularly high in coastal urban areas due to economic and demographic development. This is where the economic loss risk is likely to increase due to population growth and economic growth. This will expose more people and economic assets to natural hazards. This is particularly true for developing nations. With a loss potential of 22 percent by 2015, Tokyo ranges at the lower end compared to an 88 percent loss potential for cities like Shanghai in China and Jakarta in Indonesia.[47] Although the overall loss due to a potential future increase in typhoons in Japan in the year 2085 is a small percentage of the Japanese overall economy, it is only one part of the overall loss figures. First-order physical losses are likely to increase also, and they make up a much larger proportion of the total overall losses.

Another likely outcome of climate change is a northward shift of tropical cyclones. Figure 3.2 shows the expected number of hours (i.e., downtime) that the Japanese economy is likely to lose in the future due to increased tropical cyclone intensity. By comparing the climate change scenario with the control scenario, it becomes clear how future typhoons will shift northward. Although the typhoon tracks are not altered, increased storm intensity results into a poleward shift.[48]

Therefore, although the simulation does not alter the tracks of the tropical cyclones that reach Japan, as it alters their intensity it does increase the frequency of "typhoon grade" cyclones reaching the northern parts of Japan. This is a very important effect, since it results in increased downtime in the northern areas of Japan. The physical explanation of this effect is that an increase in surface sea temperatures would cause tropical cyclones to keep their strength over a longer distance as they travel polewards, resulting in an impact on higher latitude regions. As a consequence, northern areas of Japan previously not experiencing strong typhoons will be harder hit in the future, resulting in greater economic loss. This will affect major Japanese cities such as Tokyo, Yokohama, Nagoya, Kobe, and Osaka.

Climate models form an important tool to investigate the potential change in tropical cyclones. They contain hypotheses relating to how the climate system works, and yield different results depending on these assumptions. It must be understood that at present there is a large overall uncertainty in future changes in tropical cyclone frequency as projected by climate models with future greenhouse gas concentrations as emphasized in the IPCC *Fourth Assessment Report*.[49]

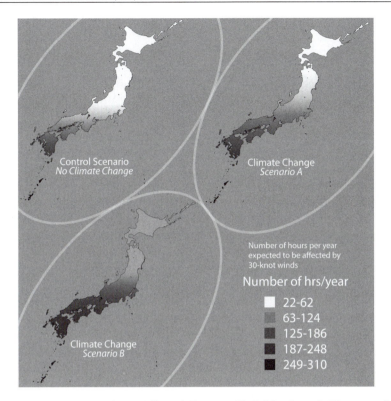

Figure 3.2 Japan's Typhoon-Affected Regions Shift Northward. The map illustrates the question of whether stronger storms mean greater losses. If CO_2 concentrations in the atmosphere continue to double by the year 2085, we can expect more typhoons reaching north with impacts for economic productivity. Thirty-knot or stronger winds already lead to a disruption of human activities. Consequently and as shown in Japan, highly populated areas such as Tokyo in higher latitudes will see more typhoons and subsequent time loss in the future if CO_2 emissions are not reduced drastically (Scenarios A and B are based on different parameters).

Source: Esteban, Miguel, "Wilder Typhoons May Mean Bigger Yen Losses," OurWorld 2.0, Media Studio, United Nations University, Tokyo (2009), www. ourworld.unu.edu.

Other uncertainties relate to the design of computational models. For example, relying on historical tropical cyclones makes it difficult to predict what future changes in global climate will have on cyclone tracks or frequencies. Hence, it does not allow for the prediction of events that are significantly different from those of the past. Also, some models rely on the assumption that larger maximum wind speeds correlate with larger tropical cyclones. Scientists have not clearly established this point in the case of large typhoons.[50]

What Does This All Mean for Human Security?

Although in Japan, storms hardly claim lives, tropical cyclones have the potential to seriously disrupt the economy, an economy that provides its citizens with educational, health, security, and infrastructural services, all essential to ensure a high level of human security. What is more, the previous analysis is human-centered, estimating the hours each person is likely to lose, before aggregating the loss for the entire economy.

The scientific evidence of climate change being a serious economic challenge is convincing. The Stern report claims, "the overall costs and risks of climate change will be equivalent to losing at least 5 percent of global GDP each year, now and forever. If a wider range of risks and impacts is taken into account, the estimates of damage could rise to 20 percent of GDP or more."[51] The projection of the economic loss in overall productivity in Japan is an attempt to move from the broader approach followed by the Stern report into a more detailed assessment of the overall cost to a specific country. Consequently, climate change adaptation will be essential for the future growth of the Japanese economy.

Although the cost of dealing with physical damage—or so-called first-order loss—of a hypothetical future of more intense tropical cyclones is not included in the previous analysis, it measures the economic non-first-order loss in Japan due to a potential future increase in tropical cyclone intensity. A 0.15 percent reduction in overall GDP in the year 2085 may sound minor, but we must add to this figure the physical, direct costs of a potential future increase in tropical cyclone activity.

The damage to infrastructure due to a future increase in tropical cyclones will be severe because infrastructure damage costs increase substantially even in the event of a small increase in sea surface temperatures.[52] The reason is two-fold: First, maximum wind speeds increase exponentially by about 15–20 percent for a 3° Celsius increase in tropical sea surface temperature,[53] and second, physical damage costs increase exponentially as a function of wind speeds.[54] A 25 percent increase in maximum wind speeds can lead to an almost seven-fold increase in building damages.[55] Along these lines, research estimates that with a doubling of CO_2 emissions in the atmosphere the destructive potential will increase by 40 to 50 percent.[56] Most important, increasing wind speeds and infrastructure damage are nonlinear events, with the largest loss event expected in the developed world. However, with growing income and industrialization, former developing countries like China will have more and higher-value infrastructure at risk.

While urban populations in coastal settings grow and prosper, the loss potential is increasing steadily.[57] This development can reach a point where the costs of natural hazards may potentially outpace economic growth. Although, due to uncertainties associated with climate changes, we should view the results with caution, they can provide a sense of the magnitude of the possible costs of climate change.

Tropical Cyclone Hotspots: Evidence from Haiti and the Dominican Republic

The study on Japan provides a detailed account of the cost of failing to mitigate climate change. Though the economic damage is likely to increase, the implications for human security are less significant. Notwithstanding, the predicted future increase in tropical cyclone intensity is more likely to have serious human security implications in low-income countries with growing populations and incomes. What other countries and regions are most likely to be affected by a potential increase in the frequency and intensity of natural disasters? And where do we expect severe human security implications?

To better illustrate this, I selected two developing countries fairly equally exposed to tropical cyclones, Haiti and the Dominican Republic. Especially in Haiti, flooding and landslides caused by tropical cyclone activity leave thousands of people homeless and claim lives every year. While the two countries have roughly the same proportion of their population exposed to tropical cyclones, sharing the same island of Hispaniola, Haiti is on average more affected, causing greater physical and economic damage and more casualties.

One of the reasons for Haiti's greater vulnerability to natural disasters is environmental change, mainly deforestation and soil degradation. The Haitian part of Hispaniola is largely stripped of forests so that water can run off unhindered, causing the devastating flooding. Even normal rains cause flooding, largely in the informal settlements surrounded by hills. To make matters worse, as it becomes visible in Port-au-Prince, flood canals are silted and barely maintained. Without forest cover, in combination with a very hilly terrain, landslides became common. When flying from Port-au-Prince in Haiti to Santo Domingo in the Dominican Republic, this becomes obvious: There is a visible contrast between the two neighbors, the darker and greener landscape on the eastern, Dominican, side and the browner and drier land on the western, Haitian side. Satellite imagery of the border region shows a sharp delineation between land patterns. A relatively large percentage, 28 percent, of the Dominican Republic remains forested today, while only 1 percent of neighboring Haiti remains forested.[58]

Historically, Haiti had dense forest prior to European colonialism. Haiti was the richer and stronger power for more than two decades, dominating the Dominican Republic for 22 years in the nineteenth century. This changed over the course of the twentieth century, and today Haiti is the

Figure 3.3 Street Scene in Port-au-Prince, 2006. Security is of great concern in Haiti's capital city. The United Nations' blue helmets assist the government to restore peace and security. Insecurity and low economic performance weakens the country's capacity to prepare and respond to natural disasters that claim lives every year.

Photo copyright © Christian Webersik.

poorest country in the New World. The economic decline can be to some extent explained by differences in some environmental characteristics, largely deforestation. There are other, less important features, such as the West-East direction of major river basins away from the Haitian side. A larger part of the explanation, however, lies in the historical and social differences, former ties to and exploitation by the colonial powers, self-defined identity, institutions, and recent political leaders. Along these lines, Jared Diamond points out: "For anyone inclined to caricature environmental history as 'environmental determinism,' the contrasting histories of the Dominican Republic and Haiti provide a useful antidote. Yes, environmental problems do constrain human societies, but the societies' responses also make a difference. So, too, for better or for worse, do the actions and inactions of their leaders."[59]

Today, Haiti faces multiple challenges: growing population, unstable governance, the influence of outsiders, drug trafficking using Haiti as a transit point, the lack of skilled labor, and few job opportunities. In addition, the frequent occurrence of tropical cyclones coupled with political instability affect economic growth. Over the years as shown in Figure 3.4, the economy appears to be less resilient to natural disasters. Cumulative figures of annual GDP show that Haiti's economy, especially since the 1980s, neither grew nor managed to bounce back from major natural disasters. Lower economic performance will in turn affect the country's capacity to adapt to an increasing intensity of tropical cyclones. By contrast, the Dominican Republic's economy grew steadily despite regime change, a long period of ruthless authoritarian rule, and the same exposure to

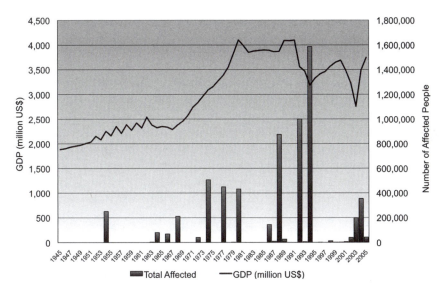

Figure 3.4 Accumulative GDP and Natural Disasters in Haiti, 1945–2005.
Haiti's gross domestic product increased steadily until the early 1980s. Since then, it has stagnated. On parallel, natural disasters affected thousands of people. Low economic growth has impeded Haiti's capacity to respond and to adapt to natural hazards. By contrast, countries such as the Dominican Republic, with a similar exposure risk to natural disasters, have been able to grow despite recurrent natural disasters (figures not shown).

Sources: EM-DAT: The OFDA/CRED International Disaster Database, Université Catholique de Louvain, Brussels, Belgium, www.em-dat.net; Oxford Latin American Economic History Database (OxLAD), Oxford University 2006; World Bank, World Development Indicators (2006).*

*GDP (million constant US$2000): Figures for 1945–1976 are from ECLAC CE (1978), GDP at factor cost. Figures for 1977–2000 are calculated with the rate of growth of GDP in constant prices from ECLAC SYLA (1984, 1987, 1993, 1996, 1997, 2002), GDP at market prices, fiscal year ending September 30. Figures for 2001–2005 are calculated from the World Bank's World Development Indicators, WDI.

hurricanes. Haiti's low coping capacity was again visible during the catastrophic earthquake on 12 January 2010 that claimed more than 220,000 lives. Low incomes and the lack of functioning institutions prevented the implementation of building codes and disaster preparedness programs that could have saved more lives. To make things worse, the unprecedented scale and magnitude of the earthquake disrupted and destroyed much of the existing infrastructure.

Haiti and the Dominican Republic are good cases to illustrate how vulnerabilities differ depending on socioeconomic circumstances. In general, climate change may act as a threat multiplier, putting constraints on societies that are already under severe stress. Other examples are Bangladesh, the Philippines, or Myanmar.

DROUGHT-RELATED DISASTERS

A major challenge to ensure human security is the predicted increase in drought frequency. Since the 1970s, droughts have increased in number.[60] Often, droughts are deadly, especially in poor countries where the sociopolitical circumstances, such as civil strife and extreme poverty, destroy coping mechanisms such as mobility or savings. More droughts will threaten food security which in Africa particularly is one of the major challenges in an era of climate change. A study undertaken jointly by the World Bank and Columbia University reveals that, worldwide, Africa has the highest risk to lose income due to drought[61] (see also Figure 6.1). Southern Africa is predicted to experience up to a 30 percent decrease in water availability for a 2° Celsius increase in global mean temperatures.[62] One of the main reasons for the risk of high economic loss is Africa's large share of agriculture in the gross national product (GNP). Natural capital, including pastureland and cropland, makes up a much larger percentage in low-income countries, compared to rich countries. Seventy-five percent of the poorest people depend on the climate-sensitive agricultural sector.[63]

Food security also depends on the type of crop, because rising carbon dioxide concentrations in the atmosphere can have positive impacts on plant growth and yield size through "carbon fertilization." Although this is the case for wheat, it appears that maize, the main staple in Africa, is less receptive to carbon fertilization, showing greater yield declines when temperatures rise, as shown in Figure 3.5. Already today, 800 million people are at risk of hunger and 4 million (almost half in Africa) die of malnutrition annually.[64] It is clear that rising temperatures and declining freshwater availability will put additional stress on societies that are least responsible for global warming but are most affected and have few resources to adapt.

Droughts, coupled with political unrest, poverty, and inequalities can have devastating effects, though some are more vulnerable than others.

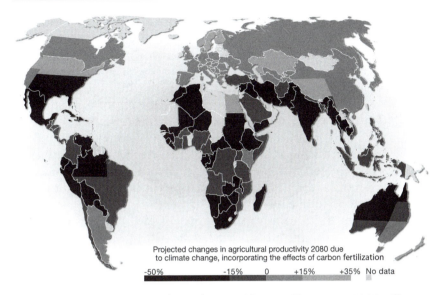

Figure 3.5 Projected Agriculture due to Climate Change in 2080. Climate change will lead to an increase in global mean temperatures, causing in some areas a wetter and in other regions a drier climate. The map shows expected changes in agricultural output given the predicted increase in temperatures, changing precipitation patterns, and carbon fertilization.

Sources: Hugo Ahlenius, "Projected Agriculture in 2080 due to Climate Change," UNEP/GRID-Arendal Maps and Graphics Library, 2008, http://maps.grida.no/go; W.R. Cline, *Global Warming and Agriculture: Impact Estimates by Country.* (Washington, DC: Peterson Institute, 2007).

For example, in 1991 to 1992, a manmade famine swept through southern Somalia. As argued earlier the famine was highly selective, affecting largely inter-riverine farmers and internally displaced persons (IDPs).[65]

A theoretical treatment of the topic can be found in Amartya Sen's work, which argues that in some famines, food was available but unobtainable by some due to unequal distribution—and not because of absolute scarcity. Sen's entitlement theory analyzes the distribution of food from the demand side and identifies the lack of access to scarce resources as a cause of hunger and death:[66] "However, starvation is a matter of some people not having enough food to eat, and not a matter of there being not enough food to eat. Although the latter can be a cause of the former, it is clearly one of many possible influences."[67] The entitlement approach tackles the problem of food supply from the demand side in contrast to the neo-Malthusian debate over "too little food for too many people."[68] People acquire food by different means. Peasants produce food for subsistence on their own land and may sell surplus in production. In contrast, workers use part of their salary to acquire food. Supply and demand set the price for each good in the marketplace—regulated by or liberated from interventionist measures of the state. Command over food is at the heart of Amartya Sen's entitlement theory. Sen calls the struggle to establish command over food the "acquirement problem."[69] According to Sen, neo-Malthusian pessimism has been disproved by an overall increase of food output per head.[70] There are problems with the entitlement approach, which tends to be ahistorical and underestimates the path-dependent development of societal structures leading to vulnerability or social resilience.[71] Nevertheless, it points to a critique of global environmental change as an absolute concept and hence potentially influences debates about the explanations of migration due to global environmental change.

Returning to the Somalia famine in August 1992, the monthly death rate in the southern Somali city of Baidoa was 3,224—104 per day. In September, the figure increased to 5,979 people, or almost 200 per day.[72] The coupling of the 1990 to 1991 drought with the civil war led to the deaths of an estimated 240,000 to 280,000 people.[73] Yet, the impact of the drought was not the most important factor causing the Somali humanitarian crisis.[74] One report from February 1992 reads: "The current famine that threatens Mogadishu and south-central Somalia is radically different in origin and impact. Drought has played only a minor role, and the main victims are poor townspeople, farmers, and rural laborers. Pastoralists are, at present, less affected."[75] Harvests were plundered and disrupted, land lay idle, and the export trade declined. This led to a decrease in domestic production of approximately 40 percent in 1991.[76] A war economy developed with militiamen selling goods at local markets at higher than prewar prices. Thousands died, and an estimated 2 million Somalis had to flee their homes.[77]

This example suggests that climate change can act as a "threat multiplier," but civil strife and poverty remain as greater challenges. Climate change and drought must be put in context, and as argued earlier, a direct link between drought and conflict seems non-existing.

The Emergence of "Water Wars"?

Another consequence of an increase in predicted drought frequency is that river flows change or become disrupted. Water scarcity can affect the domestic economic performance and sociopolitical stability of agrarian countries, but it can also impinge on international security. "Water wars" is a frequent topic in the literature. Israelis are accused of depleting the underground water resources they share with their Arab neighbors. The occupation of the Golan Heights is of strategic importance for Israel to secure the country's demand for fresh water. Another example is the dependence of Egypt on the headwaters of the river Nile. The Nile provides

95 percent of Egypt's fresh water, making the country vulnerable to any disruptions in upstream river flow.[78] History tells us that countries are more likely to cooperate on shared freshwater resources than go to war. The last international water war occurred 5,000 years ago in Sumeria.[79] Careful statistical analysis confirms this trend, although sharing the same river basin may cause some small-scale conflict.[80] However, cooperation between countries that share a river basin is the most likely scenario.[81] A good example is India and Pakistan clashing over the province of Kashmir, but they cooperate in sharing the same river, the Indus.[82]

More frequent droughts will also impact on available water for agriculture and hydroelectric power stations. Already today, agriculture uses the largest share, 70 percent, of total surface water. Industry uses only 20 percent; the remaining 10 percent goes to households.[83] The ratio may be slightly different in countries with little irrigated agriculture. This illustrates how dependent we are on freshwater resources to provide enough food and energy for a steadily growing world population.

CONCLUSION

As demonstrated in this chapter, climate change will affect all regions, North and South, rich and poor. This is an important message, moving beyond the perception that climate change mitigation is "just" an ethical debate. Climate change, as the most recent science shows, becomes incrementally a tangible threat to human well-being, in both developing (in terms of human security) and developed countries (in terms of economic security). Science shows that climate change will put constraints on health systems, infrastructure, economies, and ecosystems.[84] This chapter set out to examine the security implications of one of the consequences of climate change, the impact of natural disasters, mainly tropical cyclones and droughts.

Tropical cyclones typically affect coastal, low-lying settlements in Asia, the Pacific, and the Caribbean. Japan, Haiti, and the Dominican Republic are among those countries affected, but with different outcomes for human security. In Japan, an expected increase in the intensity of tropical cyclones will lead to additional costs, measured at 0.15 percent of annual GDP in 2085 or approximately US$60 per capita. These costs do not include the physical damage to infrastructure that would occur in addition to lost productivity. Japan's port industry is especially vulnerable to an increase in tropical cyclones because stronger storms can lead to more downtime in port infrastructure.[85] Along these lines, Esteban and others argue that "if the intensity of typhoons increases, port downtime alone could lead to a 3.4 percent decrease in Japanese GDP if an additional investment is not made of up to 127.9 billion Japanese Yen (more than 1.3 billion EUR) in port infrastructure by the year 2085."[86] A similar study projects 2085 GDP losses at 0.09 percent for South Korea and 0.7 percent for Taiwan. In Taiwan, the indirect loss increase could be greater than 50 percent. This is due to the concentration of population and economic activity in the northern part of the country that is prone to intense typhoon activity.[87] In particular, countries with growing populations and growing incomes, such as the Philippines, are at great risk to lose more income and lives due to a likely increase in tropical cyclone intensity.[88] This also applies to Haiti and the Dominican Republic in the Caribbean—both countries exposed to natural hazards. Haiti is already weakened through political instability, massive deforestation, high unemployment, and low economic productivity. To make things worse, the devastating 2010 earthquake has destroyed much of the existing infrastructure to cope with natural disasters. The capacity to withstand and cope with more hurricanes in the future will not only result in greater economic damage but will also claim more lives.

Droughts are likely to affect people who depend on drylands in Africa, Asia, and Australia. In countries that have climate-sensitive economies, droughts will negatively impact on economic performance. In addition, civil war literature shows that negative economic growth can elevate the risk of political instability as discussed in chapter 2. The risk of losing economic income due to drought is highest in sub-Saharan Africa (see Figure 6.1 in chapter 6). But other explanatory factors remain important in explaining violent outcomes, such as political exclusion along ethnic lines, high infant mortality rates, being close to international borders, and high population densities.[89] Even where violent unrest is avoidable, the implications for Africa's food security are serious. Malnutrition is claiming thousands of lives every year, and more droughts will most likely increase that number. Although some regions are becoming wetter and others drier, climatic variability will dramatically increase, with impacts on soil fertility, biodiversity, and health.

Droughts can also involve regions and countries that are not directly affected. Changing water flows can threaten livelihoods further downstream, although it is unlikely that "water wars" will emerge. Solutions for adapting to more frequent droughts include drought-resistant crops, increased dam capacity, diversification of the domestic economy, and drought insurance schemes, to mention just a few.

CHAPTER FOUR

Migration as an Outcome

Global climate is approaching critical tipping points that could lead to loss of all summer sea ice in the Arctic with detrimental effects on indigenous people and wildlife, initiation of ice sheet disintegration in West Antarctica and Greenland with progressive, unstoppable global sea level rise.

James E. Hansen addressing Japan's Prime Minister Yasuo
Fukuda in an open letter, July 3, 2008

INTRODUCTION

History tells us that humans are perfectly capable of adapting to a changing environment. Our survival during past ice ages is proof of our great adaptive capacity. Climate change will happen—and if unabated—with potentially catastrophic consequences. More extreme weather events, sea-level rise, and a hotter and drier climate are some of the predicted outcomes seriously affecting people's choice of where to live on a much more crowded planet. At the time of writing in September 2009, world population is at 6.79 billion and on the rise.[1] In the past, people had the possibility to move to less populated regions when faced by environmental change but today, population densities have increased dramatically and arable land has become more limited. In addition, large parts of the Earth's fertile land became eroded and hence unsuitable for agriculture.

Cross-border streams of "climate migrants" or "environmental refugees" caused by tropical cyclones, associated flooding and landslides, droughts, and sea-level rise could trigger resource competition with violent outcomes in the receiving country or region. But can these claims be substantiated?

Before turning to the evidence, it is worth considering some of the methodological constraints of this research. There are several reasons why it is difficult to measure the impact of climate change on human migration. One reason is the application of different methodologies and scales.

Climate studies use localized data with a high resolution in space and time, available through satellite imagery and gauge data. By contrast, migration studies often employ data sets that are only available on the country level and reflect changes for each year. The estimates of how many people move from one country to another often are rough estimates with a large margin of error. There is also very little information about the migration path of each individual, and whether the migration remains internal or becomes cross-border. It is therefore hard to link two quite different data sets on climate systems and migration for statistical purposes. In addition, there is a temporal concern: climate data are recorded on a very frequent scale, often monthly or even daily, making these data very dynamic, whereas migration data tend to be more static, collected often on an ad hoc basis and less frequently. Moreover, in climatic modeling, probabilistic forecasting and scenario studies are very common, something that is rare in human migration research. Raleigh and Jordan have recently raised these three reasons why climate change and migration are hard to link: "Problems in equating climate change and human migration research include scalar mismatches . . . temporal mismatches . . . and the treatment of forecasting."[2]

In addition, there are uncertainties involved in how the Earth's system responds to anthropogenic greenhouse gas forcing. There are complex feedbacks and thresholds at work whose underlying mechanisms are still poorly understood. Moreover, nonlinear tipping points are adding to this complexity, making it difficult to model certain outcomes.[3] Likewise, the human system is as complex as the natural system. It is more difficult to anticipate long-term policy responses or human individual and collective behavior in dealing with the outcomes of climate change. It is hard to predict economic growth or demographic long-term developments. Even though the degree of change in the climate systems is uncertain, there is scientific consensus that some of the outcomes of climate change, such as sea-level rise or biodiversity loss, are inevitable.

I have divided this chapter into seven parts. First, I list the natural hazards relevant for environmentally induced migration. A discussion of the terminology is followed by an analysis of identifying those most vulnerable to natural hazards, where they live, and how many are affected, with a focus on urbanization. The fourth section examines the impact of tropical cyclones, droughts, floods, and sea-level rise and the implications for migration. I do not discuss in detail technological accidents, expropriations as a result of anthropogenic disruptions (such as large hydropower projects), or environmental deterioration; their links to climate change are nonexistent or difficult to support with evidence. Then follows an assessment of the conflict potential of climate-induced migration and the role of the state in dealing with migration. The chapter concludes with a hotspot analysis in Southeast Asia.

METEOROLOGICAL HAZARDS, CLIMATE CHANGE, AND MIGRATION

The adverse consequences of climate change are multiple. The physical impact, the temporal and geographical scale, and the potential for migration differ enormously. The *Fourth Assessment Report* of the Intergovernmental Panel on Climate Change (AR4) has stated that it is likely to very likely[4] that climate change will lead to warmer and more frequent hot days and nights (virtually certain), more and heavier precipitation events (very likely), increased risk of drought (likely), increase in tropical cyclone activity (likely), and increased incidence of extreme high sea level (likely).[5] By taking these scenarios into consideration, droughts, floods, tropical cyclones, and sea-level rise will be most likely the main drivers of environmentally induced migration. In the AR4, however, migration is addressed as a consequence of only two climate-related natural hazards: tropical cyclones and droughts.

Droughts and famines are caused by recurrent dry months. They are triggered by a lack of precipitation. According to the Emergency Events Database (EM-DAT), "a drought is an extended period of time characterized by a deficiency in a region's water supply that is the result of constantly below average precipitation."[6] Droughts affect inland navigation, hydropower plants, and freshwater availability. Hence, droughts lead to declines in economic production (especially in the subsistence sector). Yet, the level of vulnerability to drought depends on many factors, such as land tenure, coping strategies, livelihoods, government policies, and social networks. Experts therefore argue that famines are often manmade disasters. Drought is a necessary factor but hardly sufficient to lead to a famine. Warfare, intergroup inequalities, and population pressures can weaken coping mechanisms and eventually culminate in a famine that kills.[7]

A flood is characterized by a significant rise of the water level in rivers, reservoirs, or coastal regions. Flood risks increase with deforestation, increased precipitation, snowmelt, and urbanization. People are vulnerable to floods depending on their assets, their level of preparedness, and government support to cope with flooding.

A tropical cyclone is

> a non frontal storm system that is characterized by a low pressure center, spiral rain bands and strong winds. Usually it originates over tropical or subtropical waters and rotates clockwise in the southern hemisphere and counter-clockwise in the northern hemisphere. The system is fueled by heat released when moist air rises and the water vapor it contains condenses.[8]

Tropical cyclones are responsible for considerable death and destruction along coastal regions but there is uncertainty about the impact of climate change on tropical cyclone activity. Estimates calculate a 5 to 10 percent increase in peak intensity and a 20 to 30 percent increase in precipitation between 1900 and 2100.[9] Tropical cyclones have the potential to lead to distress migration through associated flooding and landslides, though typically, people tend to return to the disaster site to rebuild their homes.

Scientists project sea-level rise to range between 0.09 and 0.88 meters between preindustrial levels in 1900 and 2100.[10] Migration will be very gradual, though there are some short-term environmental impacts with migration potential: Flood frequencies in coastal plains, erosion, and salt-water intrusion jeopardizing coastal subsistence livelihoods. Particularly vulnerable to sea-level rise are small island states and atolls, most notably, Kiribati, the Maldives, the Marshall Islands, Tokelau, and Tuvalu.[11] Most of these islands consist of low-lying atolls surrounded and protected by coral reefs.

ENVIRONMENTAL REFUGEES OR ENVIRONMENTAL MIGRANTS?

When people leave their homes due to tropical cyclones, sea-level rise, or environmental degradation, should we call them environmental migrants, environmental refugees, or simply refugees? The type, origin, length, and intensity of the environmental disruption are all important in shaping the type of migrant or refugee.[12] The literature on environmental refugees is large and growing.[13] Norman Myers has argued that environmental refugees will soon become the largest category of involuntary migration, a notion that has been contested.[14] The term *environmental refugees* was defined by Essam El-Hinnawi more than 20 years ago as

> those people who have been forced to leave their traditional habitat, temporarily or permanently, because of a marked environmental disruption (natural and/or triggered by people) that jeopardized their existence and/or seriously affected the quality of their life [*sic*]. By "environmental disruption" in this definition is meant any physical, chemical, and/or biological changes in the ecosystem (or resource base) that render it, temporarily or permanently, unsuitable to support human life.[15]

This definition is quite broad and does not distinguish whether people are forced to migrate because of gradual sea-level rise or as a result of a sudden flash flood in the Himalayan mountains. Whether or not using the qualifier "environmental," the main distinction between migrants and refugees is whether the move is voluntary or forced. According to the 1951 United Nations Convention on Refugees and its 1967 Amendment, a refugee is a person who:

> . . . owing to well-founded fear of being persecuted for reasons of race, religion, nationality, membership of a particular social group or political

opinion, is outside the country of his [*sic*] nationality and is unable or, owing to such fear, is unwilling to avail himself of the protection of that country.[16]

As the fear of persecution and hence a tangible perpetrator (often the state of origin) is the missing element in forced migration caused by climate change, lawyers and scholars with an interest in international law and security have been cautious to adopt the term *environmental refugee*. However, anthropogenic climate change contains some elements of perpetration and victimization, since those affected by climate change are often not those responsible for substantial greenhouse gas emissions. In addition, most climate change impacts, such as sea-level rise, will be global in origin but local in their impacts. It is likely that as a consequence people will be displaced internally without being "outside the country." This adds another dimension to the discussion of internally displaced persons (IDPs).

To avoid any legal implications of the term *refugee* it is helpful to accept a continuum from voluntary to forced, environmentally induced migration.[17] People who face a natural hazard like flooding must flee to avoid death and destruction. Some evacuate—sometimes against their will—without control over their relocation. Some anticipate a heightened natural hazards risk, and as a consequence decide to move before the situation deteriorates and they are forced to relocate (as for instance in the case of sea-level rise).

To incorporate all the above situations, it is useful to include the following criteria to define environmentally induced migration: The origin of the environmental disruption (natural or technological), its duration (acute or gradual), and whether migration was a planned outcome of the environmental disruption (intentional or not).[18] By applying these criteria, it is possible to accept the term *climate migrants* (for voluntary migration), *climate refugees* (for forced, cross-border migration), and *climate IDPs* (for internal displacement) when anthropogenic interference can be linked to the origin of the environmental disruption.

VULNERABLE POPULATIONS AND METEOROLOGICAL HAZARDS

Identifying vulnerable populations is the result of a shifting debate. Earlier IPCC reports emphasized migration as a direct outcome of coastal flooding or sea-level rise, whereas the most recent report puts a stronger emphasis on vulnerability and adaptive capacity.[19] This shifting framework takes the emphasis away from the pure physical risk and focuses more on the social and economic ability to cope with a changing environment. Low incomes prevent people from adapting to a greater flood or drought risk. According to the most recent assessments, the people most affected by climate change impacts, such as sea-level rise and coastal flooding, are the urban poor. There are multiple factors at play: on the one hand socio-economic factors, including the type of political institutions, gender, income, educational attainment, age, and ethnic and religious marginalization. On the other hand, there are also physical factors that shape vulnerability. These are location, infrastructure (sea defenses, flood controls, reservoirs, and dams) and geographic exposure to natural hazards. A mix of these factors determines the level of an individual's or household's vulnerability. Decisions to migrate depend on the level of vulnerability and explain why some people do not even have the choice to migrate in the aftermath of a severe drought or flood.

A differentiated approach suggests that the costs and benefits (if existent) of climate change are not equally distributed among the agents and victims of environmental change. On the international level, countries like Nepal and Haiti are least responsible for but most affected by global environmental change. Even within those countries, the costs of environmental change are distributed unevenly. In addition to human vulnerabilities, countries like Nepal or Haiti have weak and unstable governments, a history of armed conflict, corrupt officials, and high rates of unemployment and levels of inequalities. The sudden onset of a natural hazard can further weaken the economic performance of a country, as seen in Haiti. Although there is little evidence that tropical cyclones affect political stability in Haiti, a combination of political instability and the occurrence of tropical cyclones have reduced Haiti's income over the past 20 years.[20] By contrast, rich countries can afford insurance against flooding and storm damage. In the United States, companies can insure themselves against physical damage and even business interruptions caused by natural hazards.

This shows how vulnerability is often linked to income. Adaptation to climate change is costly, and an increase in extreme weather events is likely to destroy property and agricultural land. For instance, an increase in extreme weather events, such as tropical cyclone activity, will cause more flooding in coastal and low-lying areas. Floods have great damage potential: The economic loss due to floods is greater than that of any other natural hazard. A World Bank report states that high-risk flood areas (areas with the greatest mortality rates) cover a larger land area and affect a greater number of people globally than all other hazards (tropical cyclones, droughts, earthquakes, volcanoes, and landslides) combined.[21] Low-lying, coastal areas are especially vulnerable to flooding, such as the Asian mega-deltas of the Yangtze River or the Ganges. The IPCC *Fourth Assessment Report* estimates that the global proportion of this vulnerable population reaches a staggering 20 percent.[22] Although an increase in tropical cyclone intensity and associated flooding is less likely to lead to permanent displacement, it will trigger temporary population movements.

By contrast, the predicted sea-level rise of 0.09 to 0.88 meters by 2100 will result in more permanent migrations from low-lying coastal areas and low-lying islands to other, better-protected regions. Sea-level rise and storm surges also jeopardize people's livelihoods. Warmer oceans can bleach and destroy coral systems, ecosystems on which subsistence fishing depends. Saltwater intrusion poses another problem. It reduces freshwater availability for agricultural and household needs. The loss of coastal wetlands and mangrove forests will negatively affect coastal fishing communities, especially in poor countries. Already in 2004, 70 percent of all coral reefs were destroyed, in critical condition, or threatened, according to a report published by the United Nations Environment Programme.[23]

Migration is only one of the strategies for households and individuals to cope with natural hazards. Raleigh and Jordan identify three main strategies when living in uncertain ecological climates with a high natural hazard risk. These are "diversification of livelihood, consolidation of savings into incontestable forms and social investment (for example, migration)."[24] The most common reaction to a sudden hazard is the liquidation of savings, in addition to service and labor movements (for example, migration). Distress migration only occurs when natural hazards exhaust household savings and destroy the ability to cope with the natural hazards risk.

Regional Dimensions of Vulnerable Populations

It is worth asking where the most vulnerable populations live in order to estimate the percentage of people affected by natural hazards. According to a study commissioned by the United Nations Population Fund, the regions at greatest risk are eastern China, the Mekong basin, the Ganges basin in India, the great lakes region in East Africa, northern Nigeria and eastern South Africa, France and Germany, the Danube basin, eastern South America, the Andes, all of Central America, and the Mississippi basin in the United States.[25] This risk relates to current data and does not take into consideration the effects of climate change. In terms of the regional distribution of people affected by natural hazards, recorded by the Emergency Events Database (EM-DAT), the percentage of people affected (including homeless and killed) is highest in the Caribbean (5 percent) and Polynesia (15 percent), though more people in total are affected in East Asia, South Asia, and Southeast Asia. In terms of frequency, the most common natural disasters recorded between 1970 and 2004 are floods with 45 percent and windstorms with 37 percent, whereas drought frequency is at 5 percent.

Asia is especially affected. Between 1994 and 2004, about two-thirds of all 1,562 recorded floods and half of the 120,000 people killed lived in Asia. A staggering 98 percent of the 2 million people affected by flood disasters live in Asia.[26] A climate change scenario—by assuming a one-meter sea-level rise—confirms this trend, indicating the greatest migration potential in Asia. Figure 4.1 shows that more than 100 million people will be affected in Asia by a one-meter sea-level rise. With rising incomes, Asia is also likely to experience the greatest economic loss of up to US$400 billion. These, of course, are rough estimates, but the trend is clear: the regions that must fear the greatest loss are Asia, followed by Europe.

Counting the Vulnerable

Although empirical data about migration patterns of people affected by natural hazards do not exist, the data of the EM-DAT summarize the percentage of people affected by natural disasters.[27] Being affected, of course, does not translate directly into migration, but it gives an indication of the magnitude of the problem. EM-DAT also records the number of people made homeless, forcing people to move elsewhere. The data show that drought, while not the most common disaster, affects the largest percentage, on average 10 percent, of a country's population. This percentage is

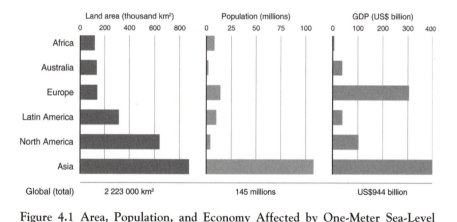

Figure 4.1 Area, Population, and Economy Affected by One-Meter Sea-Level Rise. Sea-level rise will largely affect people living in the Asian mega-deltas, most notably the Mekong and the Ganges-Brahmaputra deltas. Worldwide, 75 percent of all people affected live in deltas. With growing economic income in middle-income countries, GDP losses will be even greater. In addition, growing population densities will also increase the number of people affected. The estimates shown are based on contemporary figures, and the economic estimates represent the annual share of the domestic economy in each zone, in GDP (market exchange rates).

Sources: Hugo Ahlenius, "Population, Area and Economy Affected by a 1 m Sea Level Rise (Global And Regional Estimates, Based on Today's Situation)," UNEP/GRID-Arendal Maps and Graphics Library, 2007, http://maps.grida.no/go; D. Anthoff, R.J. Nicholls, R.S.J. Tol, and A.T. Vafeidis, "Global and Regional Exposure to Large Rises in Sea-Level: A Sensitivity Analysis," Working Paper 96. (Norwich: Tyndall Centre for Climate Change Research, 2006).

even higher (13 percent) in low-income countries, though they constitute only 8 percent of all countries. In some low-income countries, drought affects up to 90 percent of the total population in a given year, indicative of the climate-sensitive economies of these countries.

Sudden-onset localized natural hazards, such as extreme temperatures, landslides, or wave surges affect only a relatively small percentage of the country's average population, 1 percent. Similarly, tropical cyclones affect a much smaller proportion of the country's population, on average 2 percent. However, some events can disturb entire countries, with tropical cyclones affecting more than 90 percent of a country's population in seven cases, and floods affecting a maximum of 48 percent.

Population growth is one of the main factors of increased vulnerability to natural hazards. Coastal regions will experience especially significant population growth in the coming years. For example, the populations of two of the seven mega-deltas in Asia, the Ganges-Brahmaputra and the Mekong, are projected to increase to 189 million and 35 million by 2015, based on a 2000 baseline population of 147.4 and 28.2 million, respectively.[28] In addition to an increased flood risk, sea-level rise will cause salt-water intrusion in these regions, where a large proportion of the population depends on subsistence agriculture, as in Bangladesh.

Experts anticipate population growth largely in developing countries due to high fertility and a large proportion of the population reaching the reproductive age. Despite the fact that worldwide population growth dropped from 2 percent in the 1960s to 1.2 percent today, the absolute population growth remains high, adding 78 million people per year.[29] This will increase the number of vulnerable people. The reasons are twofold. First, developing countries have a lower adaptive capacity to deal with climate change; incomes are lower, and institutions are ill-equipped to prepare for and cope with natural disasters. Second, the population growth will increase the number of children and infants, who are particularly vulnerable to water and food shortages caused by droughts or the flooding from tropical cyclones.[30] Moreover, when rural regions become less populated, rural infrastructure is likely to deteriorate; social networks become dysfunctional, thus increasing physical vulnerability. The elderly, who are at most risk of becoming "victims of climate change," remain.[31]

In sum, it is the combination of population growth, increasing population densities, sea-level rise, an increase in tropical cyclone activity, and associated flooding that impact on people's choices to migrate, either temporarily or permanently.

Urbanization, Natural Hazards, and Climate Change

In addition to population growth, the world is experiencing a trend toward urbanization. Skilled and unskilled labor migrates to urban areas because of the perceived availability of employment opportunities. Much of this urbanization happens in coastal zones, most notably in China. China is an export-oriented economy, drawing large numbers of workers to the coastal regions. More generally, coasts offer easy access to international markets, and researchers argue elsewhere that landlocked countries have greater difficulties in developing their economies. This is the case in Africa, where the population is distributed more evenly throughout the continent. Urban settlements tend to grow and to be longer lasting. Urban infrastructure, such as ports and public transportation, is more costly and immobile, making it difficult and too expensive to move elsewhere. Moreover, urban areas attract enterprises and are nodes of economic productivity. This phenomenon shows elements of an urban path dependency.[32]

For instance in Bangladesh, 60 percent of urban growth can be attributed to rural-urban migration.[33] Yet, a large percentage of the urban migrants retain ties with their rural areas of origin and contribute to the development of rural communities. Today, we are about to pass the 50 percent ratio of urban to rural population with the urban proportion rising. More developed regions tend to have a higher urban to rural ratio at 73.2 percent in 2000, whereas less-developed regions (including Africa and Asia) are at 40.3 percent.[34] With growing wealth in Asia and parts of Africa, we can anticipate an increase in urbanization. Soon, around 2017, as Figure 4.2 demonstrates, the less-developed regions will reach the 50 percent ratio.

Sea-level rise, storm surge, and stronger tropical cyclones will largely affect coastal populations living just a few meters above (or even below) sea level. Worldwide, about 10 percent of the total world population (and 13 percent of the total world urban population) lives in this low elevation coastal zone (lower than 10 meters altitude) on only 2 percent of the total land area, based on population density estimates for 2000.[35] Almost two-thirds of all cities with more than 5 million people fall into this zone. Sixty percent of the population in this low elevation coastal zone is urban, whereas the global percentage is slightly lower, at 50 percent.[36] Given the global trend of urbanization and the 2000 estimates, this figure is probably slightly higher today. The four countries with the largest number of people living in the low elevation coastal zone are China, India, Bangladesh, and Vietnam, with large populated delta regions.[37] When ranking countries by share of population in this zone, the Bahamas, Suriname, the Netherlands, and Vietnam come first.[38] Moreover, the majority of these countries are in the developing world with lower adaptive capacity, yet with a high climate sensitivity (large subsistence sector and informal settlement areas) and exposure risk to natural hazards.

Bangladesh and Vietnam are particularly vulnerable to climate change, because they will be affected by sea-level rise and an increase in tropical cyclone activity. Bangladesh, for instance, has two-fifths of its population

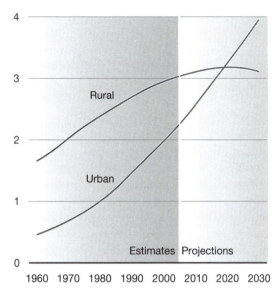

Figure 4.2 Urban Population Passing Rural Population in Less Developed Regions (in Billions). The past years showed a worldwide trend in urbanization. Also in poor regions, the urban population is growing and expected to surpass the rural population by 2017. Less a function of migration, urban growth is a result of population growth in developing countries.

Sources: Hugo Ahlenius, "Trends in Urban and Rural Populations, Less Developed Regions, 1960–2030 (estimates and projections)," 2009, United Nations Population Division, World Urbanization Prospects: The 2007 Revision Population Database; and UNEP/GRID-Arendal Maps and Graphics Library, 2009, http://maps.grida.no/go.

living in a low elevation coastal zone, with growing urban settlements there. What is more, the population in this zone is growing at a faster rate than the country's urban average.[39]

Africa, although with fewer megacities than Asia, has a higher share of its population living in cities of 100,000 to 5 million in the low elevation coastal zone below 10 meters.[40] In addition, the continent's medium-sized cities are growing at a faster rate and are poorer than those of other continents. Low urban gross domestic products reduce their adaptive capacity. Compared to cities in developing nations, they have a very small carbon footprint but they are at greater risk of climate-induced natural disasters. For instance, the city Cotonou, the economic hub of Benin with around 1 million people, has an average one-fifteenth carbon dioxide emissions compared to that in rich countries. The same applies to cities like Dhaka, Jakarta, or Mumbai.

Mumbai is another good example of a megacity—with 19 million people in 2009 and growing—prone to flooding and storm surge because large parts of the city stand on low-lying landfills. Also, new settlements (commercial, residential, and industrial) are typically low-lying and hence vulnerable to flooding. During the monsoon that starts in the middle of September, about 2,500 millimeters of rainfall is usual. Drainage systems choke with debris and waste, making flooding a common occurrence that leads to the shutdown of the commuter trains on which this economic and financial center depends. A predicted sea-level rise of 0.09 to 0.88 meters in 2100 could seriously damage Mumbai and trigger population movements from the low-lying areas toward safer areas inland. Even more problematic is the economic damage caused by sea-level rise. The Delhi-based research institute TERI estimated the economic damage at US$71 billion.[41]

The main risk for urban areas in terms of climate change is flooding and the occurrence of tropical cyclones. In August 2005, Hurricane Katrina flooded much of the city of New Orleans, leaving more than 1,800 people dead. All residents of the city were evacuated, and the worst flooding occurred in the poorer neighborhoods, largely inhabited by African Americans and the elderly. Another example was the flooding in July 2005 in Mumbai, India, that left more than 1,000 people dead and hugely affected the informal settlements of this Indian megacity of 19 million.[42] In Caracas and the North Coast of Venezuela, floods and landslides in 1999 killed 30,000 people and affected 483,000. Floods in Algiers in 2001 killed 900 and affected another 45,000 people.[43]

This is just a selection of the scale and type of impacts of natural disasters on cities. Cities are often vulnerable because of their high population densities and geographical setting: They are often located in the low elevation zone close to oceans and near the mouths of major rivers to enable commerce with the agricultural interior and international markets.

In addition to an increasing natural hazard risk, urbanization bears serious environmental consequences. Environmental and developmental factors have weakened the coastal zones' natural resilience to storm surges. Urbanization and development have led to the destruction of wetlands and mangrove forests that act as natural buffer zones. For instance, 50 percent of Bangladesh's mangrove forest has been converted or degraded today.[44] A United Nations Development Programme report published in 2006 estimates that worldwide, 35 percent of all mangrove forests are lost.[45] Moreover, rainwater drains more rapidly where developers have altered and straightened riverbeds. It is not only industrialization and rapid urbanization that destroy natural barriers. Even small rises in sea level can destroy fragile ecosystems, such as mangrove forests and sea grasses that protect agricultural land from saltwater intrusion and soil erosion. To avoid further damage and complete collapse of coastal ecosystems and to respond to the risks of flooding in coastal zones, inland migration needs to be one of the strategies of adapting to climate change.

MIGRATION AS AN OUTCOME OF CLIMATE CHANGE

People have migrated for all of human history. People move for numerous reasons including economic, social, environmental, and political factors. The literature distinguishes between so-called pull and push factors. Economic factors can be both, pulling and pushing people across borders. Individuals' or households' decisions to migrate are typically based on a rational choice to accumulate economic capital. Accordingly, differences in wage rates influence the decision to migrate. Also, individuals and households migrate to minimize the exposure to risk, both environmental and economic. But economic pull and push factors are not sufficient to explain all types of migration. Why do people of one community migrate but not another, given seemingly similar circumstances? The answer has to do with structural explanations. This theory argues that people are not only migrating as a consequence of economic preferences; factors such as human social behavior and cultural dispositions provide complementary motives.[46]

In this line of argumentation, the decision whether to migrate or not depends on the structure of the household, gender, age, educational level, and income. Migration occurs in a social context, because migrants often depend on social networks and family to move. Social ties, or in other terms social capital—the web of relationships—is an important factor determining whether and where an individual or household migrates. Another factor affecting migrants' choices is their ability to integrate into new communities at the receiving end.

Population movements are very dynamic. People respond quickly in the case of a natural disaster or political instability. Especially in the developing world, migration of one or more family members is a strategy to exploit opportunities in other regions. Remittances, money sent home by migrants, can be a significant contributor to the domestic economy. In some cases, as shown in Somalia, remittances are estimated to be far greater than the official Overseas Development Assistance (ODA).[47]

Compared to the common categories of economically or politically motivated migration, as experienced in Western Europe and North America, environmental migrants are a new phenomenon. Whereas the former category of migrants often regard themselves as victims of persecution deserving justice, the push factors for environmental migrants are less obvious. It is difficult to isolate the climate signal from socioeconomic factors. Though there is debate over whether the term *climate change victims* is appropriate, it remains difficult to identify a tangible perpetrator.

Adamo and de Sherbinin list four layers that characterize environmentally induced migration, explaining the difficulty in distinguishing environmentally induced migration from other types of migration:

1. The relation between environmental stress and population mobility is influenced by its socioeconomic context, its historical development, and its spatial differentiation.
2. Whether direct/proximate or indirect/contextual, environmental factors rarely act on their own.
3. Population mobility as response to environmental hazards is also related to people's subjective view and perception of the hazard and of their own vulnerability.
4. Finally, migration is just one among several possible responses and adaptations to environmental change.[48]

There is also a temporal dimension. Some global environmental changes that are caused or accelerated by climate change, such as sea-level rise or desertification, are relatively slow processes with long time lags. This also applies to the social sphere. Individuals or countries responsible for climate change set processes in motion that are likely to affect future generations. Therefore, any sea-level rise–induced migration will occur gradually over time.

There are also nonlinear effects to take into consideration. Sea-level rise occurs because water expands when it gets warmer. In addition, increasing snow- and ice-melt will increase sea levels. The Greenland and Antarctic ice sheets are the major contributors. Feedback loops and tipping points can accelerate the melting of the ice sheets after passing a certain threshold. The adaptation to sea-level rise will be long-term. People can relocate to higher ground nearby or migrate inland. This type of migration will be permanent.

Other, sudden-onset environmental hazards, such as tropical cyclones or floods, lead to a more rapid response. For example, Hurricane Mitch in 1998 in Honduras and Nicaragua and Hurricane Katrina in 2005 in the United States caused thousands of people to leave their homes quickly. Hurricane Mitch, which was responsible for more than 5,500 deaths and affected 1.5 million people,[49] triggered a wave of migrants within the region and to the United States. In the aftermath of the event, the United States Citizenship and Immigration Service announced the establishment of a protected persons program for those unable to return to Honduras.[50]

Other short-term environmentally induced migration or distress (forced) migration occurs when there is a constant lack of water. Especially in countries that have a large agricultural sector, droughts can lead to forced migration. In many sub-Saharan African countries, the share of agriculture as a percentage of GDP is more than one-third and much of it is rain-fed agriculture. However, in regions where droughts are common, nomads and farmers developed sophisticated systems of reciprocity to cope with environmental stress such as drought. Mobility has become a strategy to cope with the harsh environment. In pastoral communities, herders sell off their livestock to offset droughts or male household members migrate to urban areas in search of wage labor. When civil warfare or national borders prevent pastoral communities from migrating, they become vulnerable to droughts. Famine in Africa is often the result of malfunctioning coping mechanisms rather than a lack of rainfall. Research in Burkina Faso demonstrates that land degradation is a stronger factor to explain migration behavior than sporadic climate-related events such as droughts.[51]

Floods are also often responsible for short-term migration. It is a common coping mechanism to deal with environmental hazards. For example, in Bangladesh floods are important for agricultural production. When rivers change their courses, people are displaced and must move elsewhere, a common occurrence in the floodplains of Bangladesh. Typically, households relocate a short distance from their former homes because of the lack of financial means to move further, the presence of kin, and the hope that the eroded land will reemerge and can be reclaimed.[52] In this context, migration becomes a household strategy. In other cases, migration is only an option for wealthier families. In Malaysia, poverty, low educational attainment, land tenure insecurity, and the lack of government support in disaster preparedness and disaster mitigation programs are all factors that often prevent migration.[53] Consequently, poor residents often do not have the choice to move elsewhere.

A response to sudden-onset natural disasters often includes evacuation to a safer location within a short distance. It is worth noting that permanent migration is a rare outcome. A very small percentage, less than 1 percent, of the country's population becomes homeless; this applies to all disaster types.[54] In some cases, as seen in the aftermath of Hurricane Katrina, people relocate permanently. Permanent displacement is rather rare and limited to places nearby. In few cases, for instance following Hurricane Mitch, displacement was international.

One further source of mass migration could be climate change mitigation efforts, such as large-scale adaptation works (for example, water transfer schemes or flood and sea defenses). China, for instance, has planned and is building large dams to transfer water from the humid and water-abundant South to the drier North; this has already displaced millions of people.[55]

As a logical outcome, migration results from natural hazards across a continuum from voluntary to forced, and the type of migration depends inter alia on the context, the hazard type, and household characteristics such as income and educational attainment.[56] Migration induced by environmental hazards often results in the relocation of poor, rural communities to urban areas, as seen in China and Bangladesh. For example, in Bangladesh the numbers of beggars and homeless people often increase in the aftermath of droughts and floods.[57] During droughts in Mali, people migrate to Sahel cities to offset their loss of income. Similar internal migration has occurred in Ethiopia and Somalia. Migrants are often welcomed as cheap laborers, especially when demand for labor exceeds supply. Migration due to drought is less common in India and Bangladesh, possibly due to the differences in livelihoods, while populations in the Sahel and East Africa have large pastoral communities whose assets are highly mobile.

MIGRATION AND SECURITY

The most likely effect of climate change on security is on individual or human security. Climate-induced forced migration is likely to disrupt livelihoods and further impoverish vulnerable populations, with impacts on food security and structural changes of the labor markets. Making the link between migration induced by climate change and security in terms of violent conflict is more problematic. However, evidence suggests that there is a relationship between civil conflict and refugee influx. For example, wars in Mozambique, Afghanistan, Israel/Palestine, and Iraq led to millions of refugees. Such flows can alter labor markets and change the composition of ethnic and religious groups in the receiving countries or regions. For instance, the Chittagong Hill tribes who live in Bangladesh are responsible for violent conflict with the government over the influx of Bengalis from the low-lying areas. In a similar vein, Bengali migration to the northeast of India has contributed to social tension.[58] In response, the Indian government has built a 2,500-mile-long fence along the Bangladeshi border. More generally, a study claims that countries with an influx of refugees have a higher probability of experiencing conflict than countries without refugee communities. The study examined conflicts since 1950 with more than 25 battle-related deaths.[59]

To explain why a high influx of war refugees elevates the risk of conflict, it is useful to consult the civil war literature. Refugees from countries at war often have a stake in the outcome of the conflict. Sometimes, they engage in cross-border attacks on their home countries, as experienced in the great lakes region in Africa, where Hutu opposition groups formed to weaken the Tutsi-led government in Rwanda. Likewise, Tutsi refugees formed the backbone of the Tutsi-led liberation front that ended the genocide. Refugee camps are often highly politicized environments, involved in military and ideological support for one group or the other. As a result, Tamil refugees in India, Afghan refugees in Pakistan, or Rwandan refugees in the Democratic Republic of the Congo often retain strong logistical and ideological ties with combatants in their respective home countries.[60] Sometimes, the presence of militarized refugee groups provokes an invasion by the home country's military. In 1992, Israel invaded and occupied parts of Lebanon to weaken the Palestine Liberation Organization. In the case of the Democratic Republic of the Congo, the Congolese government arrested military leaders of the Tutsi militias operating on Congolese territory.

The evidence for environmental migrants/refugees is less convincing.[61] They do not bring a political agenda to the receiving areas. Similarly, there is little evidence that economically driven migrants raise the probability of political violence. Racist attacks as witnessed in Germany, and ethnic riots as seen in France in 2005 are common but short-lived and without serious political consequences. What they demonstrate, however, is the need for economic and political integration in Europe and elsewhere. Political violence does not occur in a vacuum; it very much depends on government policies and the role of the host community in the receiving area.

THE ROLE OF THE STATE

When natural disasters hit a region, there is often the call for government involvement. There are indeed examples of government intervention to deal with environmental hazards. For instance, in Ethiopia during a severe and long-lasting drought in 1984 to 1985, the government relocated tens of thousands of people from the famine-affected areas.[62] Some observers argue that this intervention was politically motivated to weaken opposition in this area. What is more, it amounted to a state-created disaster that possibly increased the number of deaths.[63] In a similar incident in Somalia, still under the leadership of the dictator Siyad Barre, several thousand nomadic people from the central regions (largely from the Abgal and Haber Gedir clans seen as opposition forces) and the Ogaden were resettled in the Brawa region at the southern Somali coast when a severe drought—still remembered as *daba dheer* or the "long tailed" among Somalis—hit the country in 1973 to 1974.

These reports lead to the question of state capacity to deal with migrant populations. Gleditsch, Nordås, and Salehyan list three levels at which migration challenges are met: The first to address the needs of migrants are local governments and host communities. If standard responses to more people, such as increased electricity needs, fresh water, housing, schooling, etc., are not met, unrest may follow. Second, on the national level, governments need to coordinate regional- and community-based responses by providing funds and ensuring legal mechanisms. Third, on the international level, the International Red Cross and Red Crescent Societies or the United Nations High Commissioner for Refugees assist national and regional agencies and governments to respond to a refugee crisis.[64] But in practice, the official response to natural disasters is often uncoordinated and late to arrive. A good example is Hurricane Katrina in August 2005, when poor coordination between local and national levels led to a higher death toll than one would expect in one of richest countries in the world.

Preparing for and responding to sudden-onset environmental hazards, such as floods and tropical cyclones, require a different set of policies than longer-term environmental challenges such as sea-level rise. For instance, in 2008 the newly elected President Gayoom of the low-lying small island state of the Maldives announced the creation of a sovereign wealth fund with money earned from tourism to purchase land in neighboring India or Australia for location of future generations to safe grounds.

MIGRATION HOTSPOTS IN SOUTHEAST ASIA

As mentioned earlier, coastal populations will face an increasing risk of storm surge and flooding. In particular, the Asian mega-deltas of the Ganges, Brahmaputra, and Mekong rivers will be at high risk of climate-induced migration. Moreover, the coastal ecosystems in Asia, such as salt marshes and mangroves, are threatened by urbanization and industrialization, reducing the size of natural buffer zones. Hydrological changes, such as the building of dams for generating electricity, affect the sediment flow toward the coasts. Reduced sediment flow not only degrades coastal ecosystems, such as mangroves; it also removes a natural barrier against storm surges. Another important coastal ecosystem, the coral reef, is under threat when sea surface temperatures rise. Coral reefs protect coasts and act as natural breakwaters. Once degradation and destruction occur, coastlines become vulnerable. In addition, saltwater intrusion into groundwater reservoirs and surface water is likely.

Demographic changes in Asia add another dimension: In addition to the estimated 200 million people exposed to tropical cyclones in Asian mega-deltas, many of the cities and countries are still growing due to population growth and coastward migration.[65] Consequently, the combination of sea-level rise, destruction of natural barriers and coastal ecosystems, and greater population density will increase the risk of greater economic and human loss in the future.

To better assess the risk and to estimate migration potential, researchers at the Economy and Environment Program for Southeast Asia have mapped the vulnerability to climate change in Southeast Asia.[66] The study used the IPCC's definition of vulnerability being a function of exposure, sensitivity, and adaptive capacity. The IPCC has defined *exposure* as "the nature and degree to which a system is exposed to significant climatic variations." *Sensitivity* is defined as "the degree to which a system is affected, either adversely or beneficially, by climate-related stimuli." Finally, *capacity* is defined as "the ability of a system to adjust to climate change (including climate variability and extremes), to moderate the potential damage from it, to take advantage of its opportunities, or to cope with its consequences."[67] To measure vulnerability, the researchers used population figures and information on five climate-related hazards: tropical cyclones, floods, landslides, droughts, and sea-level rise, and their historical records. The study also introduced a measure of biodiversity information, to measure sensitivity. A combination of socioeconomic factors (human development index, poverty, and income inequality), technology (electricity coverage and extent of irrigation), and infrastructure (road density and communication) was used to construct an index of adaptive capacity.

To obtain a vulnerability map of Southeast Asia, the study averages the normalized indicators of exposure, sensitivity, and adaptive capacity.[68] It ranks all districts and provinces according to the index and divides them into four groups, from mildly vulnerable to highly vulnerable. The analysis identifies several regions that stand out: All of the Philippines, the Mekong River Delta in Vietnam and Cambodia, the north and east of Lao People's Democratic Republic, Bangkok in Thailand, and Sumatra and Java in Indonesia. The Philippines are affected by a multitude of meteorological hazards, floods, tropical cyclones, and landslides.[69] Regions with high population densities are more vulnerable. Among those are Jakarta and Bangkok, with Jakarta being highly vulnerable to landslides and floods. Central Jakarta ranks highest in the overall vulnerability index even though it has a relatively high adaptive capacity score. This is due to the district's high exposure rate to all types of natural hazards except tropical cyclones.[70] In terms of resilience, Thailand, Malaysia, and Vietnam had a relatively high adaptive capacity, whereas Cambodia and the Lao People's Democratic Republic had low adaptive capacity.

As mentioned, vulnerability is a function of exposure to natural hazards, population density, and adaptive capacity. As a result, countries like Cambodia, with relatively low climate-related hazard risk, are vulnerable because of their low adaptive capacity. As expected, the Mekong River Delta in Vietnam and Bangkok are at high risk because they are vulnerable to climate-induced sea-level rise. The region of northern Philippines is another climate change hotspot due to its high population densities and exposure to tropical cyclones.[71] We can therefore expect climate-induced migration from these regions into other regions of the same country or neighboring countries, subject to the political and physical circumstances of cross-migratory movements. A World Bank–commissioned report that assesses the global risk of two disaster-related outcomes, mortality and economic losses, comes to a similar conclusion: Whereas Africa's mortality and economic loss risk is largely due to drought, Southeast Asia is affected by a multitude of climate-related disasters, including droughts, tropical cyclones, landslides, and floods.[72]

By identifying disaster hotspots in Southeast Asia, it is possible to foresee natural disaster risks, creating the basis for disaster preparedness and income loss prevention while avoiding forced migration. This can break the cycle of disaster event, relief, and recovery. The subnational approach allows decision makers to address localized risk factors. Organizations and government agencies dealing with disaster preparedness can anticipate where and what type of disasters might occur and plan accordingly.

CONCLUSION

Though the effects of climate change considerably amplify the risk of natural hazards, being vulnerable to climate change does not make a person a "climate migrant."[73] Exposure to an environmental hazard is only one of an array of reasons why individuals and households decide to migrate. Much of the migration (both, forced and voluntarily) caused by tropical cyclones, floods, and landslides is short-term, and affected people tend to return home in the long term.

In addition, natural hazards do not affect people equally. The urban and rural poor cannot afford to migrate in anticipation or acute danger of a natural hazard. Making matters worse is rapid urban growth, and especially the population growth in low-elevation coastal zones. The critical question remains: If the scientists' predictions come true, how many people will need to relocate or migrate elsewhere?

In conclusion, the impact of climate change on migration depends on people's vulnerability and their capacity to adapt. The fear that people who relocate to other regions or countries—either forcefully or voluntarily—trigger violence and political turmoil is largely unsubstantiated. Sudden-onset natural hazards cause most people to move temporarily, whereas sea-level rise will lead to permanent but very gradual relocation. The challenge for governments remains to facilitate integration of newcomers socially and economically. On a more individual level, sudden-onset disasters, such as tropical cyclones and floods—predicted to increase in intensity—will most certainly have serious implications for human health and well-being while undermining human security. Migration (either short-term or permanent, forced or voluntarily) may become inevitable, also as a strategy to adapt to climate change.

CHAPTER FIVE

Ripple Effects of Climate Change Mitigation

Anyone who has never made a mistake has never tried anything new.

Albert Einstein

INTRODUCTION

Globally, regionally, and locally, individuals, firms, and governments have begun to reduce emissions by using energy more efficiently or by replacing fossil fuels with renewable energy such as water, solar, or wind. But renewable energy as a percentage of total global energy use remains small. In the medium-term, we will see a continuation of the use of fossil fuels. Oil remains a critical resource to enable us to be mobile, making it difficult to substitute it with other forms of energy. Coal as one of the main suppliers of energy could be more easily substituted with "clean" energy sources. Coal is central to the climate problem, with the largest potential for future emissions. Consequently, analysts demand phasing out coal to reduce emissions drastically, as argued by the director of the NASA Goddard Institute for Space Studies and Adjunct Professor at the Columbia University Earth Institute, James Hansen.[1]

There is a range of suggestions and strategies to reduce emissions. Coal will remain a great challenge unless, as proposed by scientists, the carbon dioxide (CO_2) released during the combustion process is captured and stored. Capturing and storing or sequestering carbon in rock formations is technically feasible. To avoid compromising mobility, there are plans to replace oil with biofuels based on renewables. There are other proposals to "geoengineer" the Earth's energy system, including painting roads and buildings white or installing large reflectors in the atmosphere to change the Earth's energy system. What seems to originate from science fiction is being discussed seriously in scientific circles.

Many of the efforts to mitigate the impact of climate change by reducing greenhouse gas emissions, however, will have unintended social costs that may increase the risk of new conflicts. For example, as increasingly large areas of agricultural land are given over to land-extensive renewable energy technologies such as biofuels, global agricultural production is likely to decline while food prices may increase. Another set of human security risks relates to geoengineering projects. One example is ongoing experiments to fertilize oceans with iron to stimulate the growth of algae that can absorb CO_2. In 2009, an Indo-German research project employing this method was heavily criticized for its environmental impact.[2] Even more importantly, it was rather unsuccessful: During the experiment, predators ate much of the algae microorganisms, thus limiting the expected algae growth, and as a result, only a negligible amount of CO_2 was sequestered. Another plan is to store CO_2 in geological formations, a technology that bears a smaller risk. Even more worrisome are large geoengineering projects that interfere with the terrestrial energy cycle. The idea is to block radiation physically by adding small particles to the atmosphere or by installing reflective devices on the Earth's surface. By doing so, scientists hope to counterbalance global warming, while accepting great environmental risks. Another concern is nuclear proliferation and the risks associated with nuclear energy. When governments such as Iran use more nuclear energy to substitute for fossil fuels, this raises concerns about nuclear safety as well as the proliferation of nuclear weapons. The following gives a snapshot of the most important types of mitigation strategies and outlines their potential implications for human as well as international security.

BIOFUELS AND FOOD SECURITY

One of the controversial issues of climate change mitigation is the fuel-food conflict. Biofuels have great potential to reduce CO_2 emissions. First-generation biofuels are generated from a variety of crops, such as palm oil or rapeseed. They are used largely in the transportation sector as well as for heating and cooking. In December 2008, a pilot trial at New Zealand Air using a biofuel-gasoline mix for a Boeing 747 passenger plane was completed successfully.

Biofuels are solid, liquid, or gaseous fuels generated from relatively recently lifeless or living biomass. By contrast, fossil fuels develop from long-dead biomass. Basically, fossil fuels are incompletely decomposed organic matter. There are two ways of producing liquid and gaseous so-called first-generation biofuels. First-generation biofuels mainly utilize the seeds and fruits of the plant: One is through crops with a high sugar content, such as sugar cane, sugar beet, corn (maize), or cassava. A process of fast yeast fermentation generates ethanol (ethyl alcohol). A second way to produce biofuel is to use plants that contain vegetable oil, such as rapeseed, palm, soybean, or jatropha oil. The generated oil can either be used directly in a diesel engine, or be chemically refined to produce biodiesel.

For example, rapeseed is widely used in Europe, where it generates more oil per land area unit than other crops such as soybeans. In terms of price, however, it is hardly competitive with fossil fuels such as diesel, as growing, crushing, and refining rapeseed is costly. Most important, there is concern that a small percentage, around 3 to 5 percent, of the fertilizer nitrogen is released into the atmosphere as nitrous oxide (N_2O), a gas that is believed to be 296 times more potent than the greenhouse gas CO_2. The same applies to biofuel ethanol from corn. In addition, the growing, harvesting, and refining process also consumes fossil fuel, which needs to be incorporated into the balance of CO_2 savings. If the figures are right, the emissions released to the atmosphere can cancel the cooling effect of fossil fuel savings.[3]

Rapeseed is used for biodiesel production, whereas the majority (90 percent) of biofuel comes from bioethanol, largely produced from sugarcane and corn, as Figure 5.1 illustrates. In Asia, the most important feedstock crops are sugarcane and palm oil. In terms of efficiency, cassava has the highest bioethanol output per hectare, followed by sugarcane. Regarding biodiesel, palm oil is the most efficient, followed by jatropha and coconut. Still, the global total production of combined bioethanol and biodiesel is small. It only accounts for 1 percent of the total global transport fuel market.[4]

Figure 5.2 demonstrates how much food prices have soared in recent years as demand for feed, biofuel, and food has increased. There are many other reasons, such as global population growth, changing diets with an increase in meat consumption (as experienced in Japan[5]), droughts, storms, and floods. As a consequence, food prices jumped by 20 percent in 2007 before they peaked in June 2008 and since then dropped slightly.[6] The price of wheat has tripled; important staples for developing nations, such as rice and corn, have almost doubled since 2000.[7] The impact of soaring food prices on poor households is dramatic because those households spend more money proportionally on food than wealthier people do. While food purchase represents only 10 percent of the overall household income in the United States, it accounts for more than 60 percent of income in sub-Saharan Africa.[8] Already today, millions of people, largely children, are undernourished, and this figure may triple in sub-Saharan Africa between 1990 and 2080 if climate change continues unabated.[9] The riots over rising food prices in Somalia and Egypt are just precursors of the conflict potential that food shortages and associated food price hikes pose.

There is also an international aspect: countries such as Egypt, Cambodia, India, Indonesia, and Vietnam have banned rice export to guarantee domestic supply, causing international repercussions for food security. In

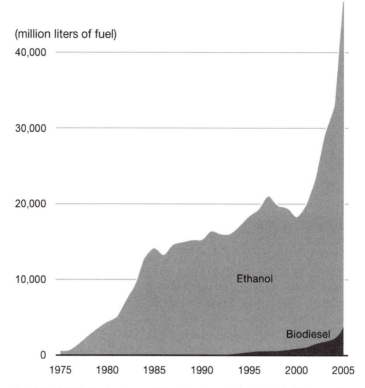

Figure 5.1 World Biofuels Annual Production, 1975–2005. The need to reduce CO_2 emissions and energy security concerns has boosted biofuel production. In the last five years, the production of biofuels has dramatically increased. The bulk of this growth comes from ethanol produced from sugarcane, corn, and soybeans.

Sources: Hugo Ahlenius, "The Production of Biodiesel and Ethanol Has Increased Substantially in Recent Years," UNEP/GRID-Arendal Maps and Graphics Library, 2009, http://maps.grida.no/go.

other countries, such as China and Argentina, governments imposed taxes or quotas on imported rice and wheat to reduce their dependency on international markets.[10] The reasoning is the protection of domestic populations, especially the vulnerable and poor, by guaranteeing them food below world prices before allowing the export of food at higher margins. On the contrary, advocates of a free-trade system argue that these protective domestic measures are not solving the food problem but rather contribute to higher food prices in other parts of the world.

By contrast, increasing food prices can also open a new window of opportunity. They can offer African and Asian farmers a chance to compete with commercial agriculture, as long as they have government support for agricultural inputs, seeds, fertilizers, and marketing infrastructure. As a researcher of the Columbia University–based Earth Institute argues: "When

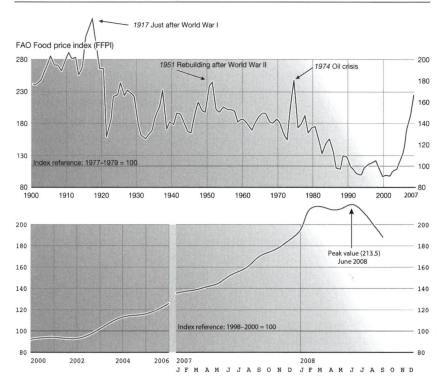

Figure 5.2 Food and Agricultural Organization Food Price Index, 1900–2008.
The food crisis and the recent price hikes are the result of low food stocks, growing demand for biofuels, and droughts and other natural disasters in crop producing regions like Australia. The crisis has led to a several-fold price increase that peaked in June 2008, driving millions of people into poverty.
Source: Hugo Ahlenius, "Trends in Food Prices," Food and Agricultural Organization of the United Nations (FAO); and UNEP/GRID-Arendal Maps and Graphics Library, 2009, http://maps.grida.no/go.

the prices are high, African governments should do all they can to increase the food production. By supporting agricultural inputs, as the government of Malawi has done with subsidy programs for seeds and fertilizers, the much needed African Green Revolution can be made operational."[11] What farmers in Africa need is food aid to a lesser extent, but rather more input support, to move from subsistence to smallholder agriculture, and from smallholder to commercial agriculture.

Other worrying aspects of growing international demand for biofuel production are the environmental impact, especially deforestation. Indonesia has been at the forefront of palm oil production. To make room for new plantations, developers razed large areas of pristine rain forest.[12] Among these are peat lands that store large amounts of CO_2, which is released to the atmosphere when peat lands are converted to plantation land. When peat land is burned, it releases especially large amounts of greenhouse gases. Deforestation and forest degradation are major contributors to global greenhouse gas emissions. The latest assessment report of the Intergovernmental Panel on Climate Change places deforestation and forest degradation as the second largest sources of CO_2 after fossil fuels, namely gas, oil, and coal. The report estimates that emissions from deforestation and forest degradation account for 17.4 percent of global total emissions.[13] This is a very large share, shedding a controversial light on the use of biofuels. Deforestation, largely of pristine forests, proceeds at an alarming rate. Most of the tropical countries in Latin America, Africa, and Southeast Asia lost forest cover between 2000 and 2005, with high rates in Indonesia, Nigeria, and Brazil, among others.[14]

A way to avoid the food-fuel conflict is the generation of biofuel from second-generation biofuels. Second-generation biofuels can be produced from a wider range of sources, largely nonfood crops including wood and agricultural residuals, municipal waste, and microalgae.[15] Still, some of these sources need pesticides and fertilizers, limiting the potential to reduce greenhouse gas emissions. Even so, the yield is much higher compared to first-generation biofuels because the entire plant can be used, not just the grains or seeds. The great potential of cellulosic feedstocks, such as aquatic organisms, may help to make second-generation biofuels a feasible alternative to fossil fuels. Of course, this option is not free from environmental impacts. There is the concern that the removal of biomass from forests may reduce biodiversity, or organic material needed to maintain soil quality may become scarce.

Today, second-generation biofuels have not yet entered commercialization. Despite some small-scale projects, such as central heating from wood residuals in Germany, it is believed that biofuel technologies will not be available in the market until 2030.[16] As mentioned, the potential is large—India, Indonesia, and China alone could produce 341 billion liters of ethanol from rice, wheat, sugarcane, and corn residuals alone.[17] There is potential to produce more from other sources, such as timber mill residuals, construction waste, organic rural and urban waste, and forests.

Another way to produce biofuels is from algae, which has great potential, but so far has proven to be difficult to commercialize. The advantage of algae is that it needs little input to yield high output. In comparison with other crops, such as soybeans, algae generate an output 30 times higher per hectare. This great leap in efficiency makes algae a promising alternative. To better illustrate this, the United States would need to cover half of its land with soybeans to satisfy the entire country's diesel needs, according to a California-based private firm.[18] By contrast, the cultivation of algae would only take up approximately 15,000 square miles, which is a bit larger than the state of Maryland.[19] One of the great advantages of cultivating algae is the possibility of growing it in salt water, thereby not competing with food crops such as soybeans or corn.

There is also great interest in bioengineering technologies to improve the features of plants to increase per-hectare yields by making them herbicide and insect resistant. The most common genetically modified crops are soybeans, followed by corn, cotton, and rapeseed. The European Union (Directive 2001/18) defines a genetically modified organism as an "organism . . . in which the genetic material has been altered in a way that does not occur naturally by mating and/or recombination." The Cartagena Protocol on Biosafety is more general, using the term *living modified organism*: a "living organism that possesses a novel combination of genetic material obtained through the use of modern biotechnology."

There is also recent research on the genetic modification of second-generation biofuel plants such as trees to increase output per hectare while improving the plant's resilience to pests and insects. But bioengineering and genetically modified plants raise a new set of environmental concerns in terms of biosafety and the risk of invasive species or weed resistance, just to name a few. For example, when weeds rather than feed crops develop resistance they may adversely affect agricultural productivity and hence food security. Another example is the risk of transferring genes from plants to microorganisms. There are other implications of genetically modified plants. One of the concerns is the risk of gene transfer to wild relatives, also called genetic contamination. This risk is small when the cultivation of genetically modified plants occurs in nonnative regions, for example, the cultivation of corn or soybeans in Europe. However, once the climate becomes warmer, corn plants have a better chance of surviving the European winter, increasing the plant's invasive potential. The dispersion of pollen is another challenge, especially where agricultural areas are close to protected areas. Genetically modified corn may generate insect-resistant pollen. When this pollen travels over long distances and lands on insect food plants, including endangered species in protected areas, it can cause damage to organisms that are crucial to the respective ecosystem.[20] Ultimately, solutions—such as genetically modified plants—that were designed to mitigate climate change may ultimately have an adverse impact on food security and biosafety.

CARBON SEQUESTRATION

Policy makers and scientists alike have great hope for large-scale geoengineering projects to capture and store atmospheric CO_2. There are two places to store CO_2: one is in the deep ocean, and the other in underground rock formations. To store CO_2 in the deep ocean, a biological option is the fertilization of the oceans with iron to stimulate phytoplankton growth that—in theory—will increase the organic carbon flux to the deep ocean. Geological storage for carbon implies the use of depleted oil or gas fields, coal layers, or deep saline reservoirs to store CO_2 permanently. Both biological and geological methods are quite controversial in their environmental risks and effectiveness. Both biological and geological methods are quite controversial in their environmental risks and effectiveness. Some of these reservoirs, however, have successfully stored gaseous substances for thousands of years.[21]

Ocean Fertilization

The idea behind ocean fertilization is that iron is the factor limiting the growth of phytoplankton (or planktonic algae).[22] When phytoplankton die and sink to the sea floor, they sequester some of the carbon taken from the atmosphere on the ocean floor.[23] Thus, proponents of ocean fertilization argue that vast amounts of carbon could be sequestered on the ocean floor by stimulating phytoplankton growth through the "fertilization" of patches of ocean with iron.

This technique ties in with the notion of the "biological pump" that removes CO_2 from the atmosphere and stores it on the deep ocean floor.[24] Carbon dioxide, converted through photosynthesis by single-celled aquatic organisms (phytoplankton) that are largely consumed by larger organisms, is released again later through respiration. Only a very small fraction of organic carbon sinks toward the ocean floor, where it is stored. Scientists hope that iron fertilization would increase the amount of phytoplankton that would sink to the ocean floor before other aquatic organisms could consume it. This would reduce the amount of CO_2 in the ocean surface water and subsequently, the atmosphere. Though phytoplankton account for only 1 percent of total photosynthetic biomass, phytoplankton are responsible for approximately half of the global organic carbon fixation. This raises hopes that commercial fertilization of the oceans can fix and store a significant proportion of atmospheric carbon.

Some ocean regions are scarce in the biologically available iron needed for the growth of phytoplankton. Thus, fertilizing the ocean with iron would trigger the growth of phytoplankton (or planktonic algae). The hypothesis follows the logic that if more iron were available in these regions, more organic carbon could be fixed by increasing organic carbon in the deep sea.[25] A famous quote by John Martin summarizes this debate: "Give me half a tanker of iron, and I will give you the next ice age."[26]

Ever since Martin's inflammatory words, scientists have carried out experiments to test his hypothesis. From 1993 to 2005, 12 oceanographic expeditions have been undertaken in the North Pacific, the Equatorial Pacific, and the Southern Ocean. Some of these areas are devoid of iron though they are rich in other nutrients. Experiments have shown that iron indeed leads to a bloom of plankton in the ocean.[27] But this technology has its staunch opponents, who claim that ocean fertilization impacts ocean ecosystems in unintended ways. The unintended environmental consequences of large-scale iron fertilization of the oceans can pose risks to human security: Large amounts of iron can potentially change and damage the entire biological food chain of the oceans, on which humans depend for food security.

As early as 2001, Chisholm, Falkowski, and Cullen published an article in Science Magazine criticizing ocean fertilization.[28] The authors provide a number of possible ripple effects of ocean fertilization, suggesting "the oceans' food webs and biogeochemical cycles would be altered in unintended ways" and that "sustained fertilization would likely result in deep ocean hypoxia or anoxia [oxygen depletion]. This would shift the microbial community toward organisms that produce greenhouse gases such as methane and nitrous oxide, with much higher warming potentials than CO_2."[29] Chisholm and colleagues also hint at the ineffectiveness of ocean fertilization as a sequestration strategy.

The enrichment of water bodies with nutrients from agricultural production and sewage can lead to eutrophication and even the complete collapse of all aerobic organisms, something humans have tried to avoid and to reverse in coastal regions and lakes. It is not possible to change the carbon cycle of the oceans without altering the coupling biogeochemical cycles that sustain ocean ecosystems. Though the potential of sequestering 15 percent of the anthropogenic CO_2 through ocean fertilization is enormous, we need to weigh risks against potential benefits. More important, the problem of permanence persists as deep ocean reservoirs are eventually re-exposed to the atmosphere through global ocean circulation currents.[30]

One of the limitations of ocean fertilization is that it fails to transfer carbon from the fast, biological carbon cycle to the slow, geological cycle. There are two different carbon cycles, working at different speeds. The fast one is biologically driven, when forests and phytoplankton act as sinks. The other, much slower cycle, is driven by volcanic outgassing of CO_2 coupled to the metamorphic weathering of silicate rocks, a process that takes millions of years.[31] Precipitation washes carbon into the ocean where it either sinks to the bottom of the ocean or is consumed and deposited on the ocean floor by marine organisms. Sediments slowly accumulate and transform over millions of years into geological formations. In fact, most terrestrial carbon is actually stored in the lithosphere in the form of inorganic carbon stored in, for example, limestone, or in organic carbon in the form of fossil fuels.[32]

When biomass converts to fossil fuels, carbon is transferred from the fast, biological cycle to the slow cycle, controlled by volcanic eruptions. By using fossil fuels, we bring carbon from the slow to the fast cycle. The creation of biological sinks, such as forests and phytoplankton in the ocean, cannot hold pace with the speed with which we are adding CO_2 to the atmosphere. Neither adding CO_2 to the deep sea through increased plankton growth nor injecting it directly into the sea sequesters CO_2 back into the slow carbon cycle from which it comes. Given these limitations and the risks of ocean fertilization in terms of ecosystem functions and food security, these types of geoengineering options need cautious deliberation.

Lohafex—An Ocean Fertilization Experiment

In March 2009, an international team of scientists mainly from India and Germany gathered in the city of Punta Arenas at the southern tip of Chile and set sail aboard the research vessel Polarstern, bringing with them six tons of dissolved iron, which they intended to dump in the Southern Ocean. This project, called Lohafex (*loha* being the Hindi word for iron and Fex standing for Fertilization Experiment) was run by the Indian National Institute of Oceanography and the German Alfred Wegener Institute for Polar and Marine Research and aimed to investigate a controversial proposed climate change mitigation scheme—ocean fertilization.

Due to mounting environmental concerns, the 9th Conference of Parties to the Convention on Biological Diversity (CBD COP 9) held in May 2008 in Bonn, Germany, issued a clear set of rules for ocean fertilization activities that

> requests Parties and urges other Governments, in accordance with the precautionary approach, to ensure that ocean fertilization activities do not take place until there is an adequate scientific basis on which to justify such activities, including assessing associated risks, and a global, transparent and effective control and regulatory mechanism is in place for these activities; with the exception of small scale scientific research studies within coastal waters.[33]

Yet despite this moratorium, and despite being suspended by the German government in January 2009, Lohafex went ahead, dumping six tons of iron into a 300-square-kilometer patch of ocean.[34] This experiment was neither small-scale, conducted in coastal waters, nor executed in the presence of any sort of regulatory mechanism—thus many have argued that it was a blatant breach of international law.[35] The environmental activist group Greenpeace mounted a campaign to stop the experiment.[36]

Nevertheless, the experiment went ahead, and yielded some interesting if frustrating results. A press release quotes Prof. Victor Smetacek: "To our surprise the iron-fertilized patch attracted large numbers of zooplankton predators belonging to the crustacean group known as amphipods."[37] This predation prevented significant algal growth and resulted in a negligible amount of CO_2 sequestered.[38] The scientists suggest that previous experiments yielding greater results involved a group of algae called diatoms, which are protected by silica shells from predators.[39] However, there is an absence of silicic acid, required for such silica shells, in the Southern

Ocean, leading the phytoplankton to become an easy lunch for the predator zooplankton.[40]

The Lohafex experiment provides an excellent example of some of the challenges of climate change mitigation strategies. It shows that unexpected and complex ecological factors can often arise from unanticipated sources, and that such strategies can face legal, ethical, and political debates equally as fierce as the parallel scientific debates.

Carbon Storage in Geological Formations

"CO_2 capture and sequestration could help save the world, if anyone's got the stomach to risk it," writes William Pentland, for *Forbes*. While recognizing that the technology has enormous potential for climate change mitigation, he cites a horrific accident at Lake Nyos, Cameroon, in 1986 as "evidence of the potential risks that have hobbled efforts to commercialize what many consider the only realistic way to satisfy the world's enormous energy needs without accelerating the pace of climate change."[41] Lake Nyos had an enormous reservoir of naturally sequestered CO_2, which leaked into the atmosphere on August 21, 1986.[42] Carbon dioxide is much heavier than oxygen or nitrogen, and it formed a cloud close to the ground that displaced all of the oxygen and caused more than 1,700 nearby residents to die of asphyxiation.[43] Decades later, with carbon capture and sequestration considered as a strategy to remove CO_2 emissions from the atmosphere, many have drawn parallels between the Lake Nyos incident and the possible dangers of carbon sequestration.

Carbon capture and storage (CCS) in its most common form, consists of separating CO_2 from the emissions of large emitters such as power plants, oil refineries, cement factories, and even highways, purifying it, compressing it into a liquid, and transporting it via a pipeline to either an underground or ocean storage site.[44] Some of the possible storage sites include depleted oil and gas reservoirs, brine-filled formations, deep unminable coal beds, the mid-depth ocean, or just above the ocean floor.[45]

Despite the persistent references to the terrifying accident in Lake Nyos, many scientists take a much more optimistic and confident approach to CCS. Benson and Surles argue that there is technological experience with each of the elements of CCS, which should reduce risks: "Almost every element of this technology is employed on an industrial scale today, for a variety of purposes. . . . Lacking is the integrated experience applied to electricity generation and cost-effective approaches to CO_2 capture."[46] Carbon dioxide is commonly used in all sorts of industrial processes. Benson and Surles argue that "because of its extensive use and production, the hazards of CO_2 are well known and routinely managed. . . . Carbon dioxide capture and transportation pose no unique risks that are not managed routinely in comparable operations."[47] With respect to storage and leaks, they acknowledge that leakage could harm groundwater and ecosystems, lead to soil acidification, and cause health and other environmental risks.[48] But they point to "analogous experience with gas and liquid injection" in other industrial processes including natural gas storage.[49] The 2005 special report by the Intergovernmental Panel on Climate Change agrees: "While there is limited experience with geological storage, closely related industrial experience and scientific knowledge could serve as a basis for appropriate risk management, including remediation."[50] It also adds that "standard well repair techniques" could stop any leakage from a storage site.[51]

In addition to leakage, storing CO_2 deep underground comes with volumetric changes and fluid pressure changes. As a result, the injection of CO_2 deep underground can alter stresses in the interior of the Earth's crust. This process can bring preexisting faults to failure, ultimately triggering earthquakes. The phenomenon of human-triggered earthquakes or geomechanical pollution has been understated in the literature and is one of the concerns of large-scale carbon sequestration.[52] Most of such human-triggered earthquakes rupture on preexisting faults close to the Earth's surface. Though there are no reports that the injection of fluids such as liquid CO_2 triggered earthquakes, there is evidence of mining activities triggering earthquakes. In 1989 in Newcastle, Australia, an earthquake killed 13 people and caused up to US\$3.5 billion damage, an event believed to be caused by deep coal mining.[53]

Today, in fact, in a limited number of sites, CCS technology is already in place. For example, a joint venture between British Petroleum and an Algerian state oil company is currently involved in "the largest operational CO_2 capture project in the world," which re-injects CO_2 emissions released during exploitation of the In Salah gas field in Algeria into the gas field. In addition to removing emissions, this process actually increases the yield from the gas field.[54]

The injection of CO_2 into former gas, oil, or salt reservoirs is a considerable low-risk option. An IPCC report and most of the academic literature seems to suggest that if proper engineering techniques are used, the risks are quite minimal.[55] Given the experience with CO_2 capture, gasification, and underground injection of gases and liquids, this option is ripe for commercialization. The challenges are more of a financial and technological nature than objections related to environmental risks or security. Although there remain a number of security-related concerns regarding CCS technologies, it seems that CCS technology has the potential to sequester 1,000 billion tons of CO_2 in brine-filled formations alone, a multiple of the global annual CO_2 emissions, reducing greenhouse emissions drastically.[56]

NON-CARBON-RELATED GEOENGINEERING

Most measures of climate change mitigation, such as replacing fossil fuels with biofuels or sequestering CO_2, aim at either avoiding or reducing the amount of CO_2 and other greenhouse gases in the atmosphere. In *Time*, Bryan Walsh points out that "the world has utterly failed to do so,"[57] and we are far from stabilizing CO_2 concentrations, especially when countries such as China, the United States, and Russia continue to rely on fossil fuels, most problematically, coal.

Scientists, including the 1995 Nobel Prize winner Paul Crutzen, have proposed to go beyond mitigation (reducing or avoiding greenhouse gas emissions) and adaptation (taking measures to live with the adverse consequences of climate change), suggesting more radical measures that are also known as geoengineering—as opposed to bioengineering measures—and aim at altering the Earth's geophysical systems. Most of these types of geoengineering measures aim at mimicking natural processes such as blocking solar radiation or changing the Earth's reflective attributes. The most important of these processes that occur naturally are volcanic eruptions. We know, for instance, that following the eruption of the volcano island of Mount Krakatau in Indonesia in 1883, the global temperature fell by an estimated $0.3°$ Celsius.[58] This cooling effect was caused by the release of volcanic aerosol clouds that reflected a portion of the incoming sunlight. In addition, aerosols also diffuse sunlight, allowing plants to photosynthesize more efficiently. This effect has resulted in increased tree growth (measuring tree rings of live and dead biomass), following the 1991 Mount Pinatubo eruption in the Philippines.[59]

Geoengineering strategies vary in their costs, methods, and potential ripple effects, but most of them are focused on the idea of solar radiation management; in other words, compensating for the increase in greenhouse gases by directly "reducing the incidence and absorption of solar radiation."[60] Many of these schemes are in the early stages of research, and involve increasing the Earth's albedo, or reflectivity, by around 1 percent, which Michael MacCracken suggests would be "about the level of effort required to offset the radiative forcing for the emissions projected for the 21st century."[61]

The scheme that has received the most attention is the increased loading of small particles into the atmosphere and stratosphere. In 2006, Crutzen proposed injecting sulfate aerosol particles into the stratosphere. The aerosol sulfate is a suspension of fine solid particles or liquid droplets in a gas. Other examples include smoke, oceanic haze, or air pollution. Alternative proposals include injecting these particles into the mesosphere or troposphere, or injecting soot instead of sulfate.

A number of these schemes, if they turned out to be financially and physically feasible, would have an enormous effect on the global climate, and have the potential to counteract quite effectively what appears to be an unavoidable increase in greenhouse gases over the next century. But it is important to note just what geoengineering means—altering the energy balance of the entire globe—and in this light not everyone is convinced that any of the geoengineering schemes are worth the risks and possible ripple effects they entail.

While the above approach is the most practical, there are others, some sounding as if they would be more useful in a science fiction film. These include reflective measures on land, such as covering roughly 70,000 square kilometers of desert land with a reflective Mylar® sheet, or installing a floating reflective structure on the ocean, which would need to be the size of a continent. Hard to imagine in the era of global warming, there are also various schemes that involve increasing or strengthening sea ice, which has a very high albedo (because it is white). Alternatively, boats could spray seawater into low-level clouds, which would increase the albedo of these clouds. Other approaches are even more science fiction-like, such as placing a solar deflector disc between the Earth and the sun. This massive reflector would need to be around 1,400 kilometers in diameter, and would require a manufacturing plant on the Moon. An alternate but similar proposal suggests launching about 10 trillion smaller deflectors to the same point in between the Earth and the Sun. Another suggestion is to place such solar shielding in near-earth orbit as opposed to geostationary orbit.

MacCracken is careful to note a number of risks, and Alan Robock of Rutgers University has written a detailed article titled "20 Reasons Why Geoengineering May Be a Bad Idea."[62] Some of these are especially relevant to security issues. For example, Robock and MacCracken both note that scientists have used the comparison of the eruption of Mount Pinatubo in the Philippines in 1991, which involved increased loading of sulfur dioxide into the atmosphere and caused significant cooling, as evidence suggests that loading sulfate aerosols into the atmosphere would be relatively harmless. Robock rejects this claim, asserting instead that volcanic eruptions often have important effects on regional climate, and can cause drought and famine, sometimes halfway around the world. MacCracken also points out that decreased direct solar radiation from sulfates would disrupt the hydrological cycle. In terms of security, this is an important consideration, because water shortages and resulting droughts influence food security. More CO_2 in the atmosphere is already leading to changes in the hydrological cycle. The double-manipulation of the composition of the Earth's atmosphere may have unintended reinforcing effects, rather than neutralizing each other.

Sulfate aerosols would also interfere with other atmospheric processes, and would be detrimental to the ozone layer and increase acid rain, according to both MacCracken and Robock. This has adverse health effects, potentially harming natural ecosystems as well as crops, and possibly exacerbating the world food crisis. Robock also brings up the question of control and management. There is obviously no world body with the mandate or authority to control the global climate, and such a drastic intentional alteration of the climate, whether undertaken by international organizations, private firms, or national governments, would presumably cause conflict. Robock suggests that conflicts would arise over "control of the thermostat" with different nations and regions wanting the global "thermostat" set at a different temperature.[63] He further claims that any geoengineering scheme that adversely affects regional or local climate (e.g., causing drought) would be in violation of the Convention on the Prohibition of Military or Any Other Hostile Use of Environmental Modification Techniques, or ENMOD of 1978.

In fact, geoengineering has a political dimension. Huge amounts of sulfur dioxide injected into the upper atmosphere to simulate the cooling effects of volcanic eruption will cross over national boundaries with significant health risks when they travel beyond the stratosphere. Jamais Cascio foresees a grim future: governments turning to geoengineering measures and technologies and using them as weapons.[64] If geoengineering enters commercialization despite the uncertainties and risks, we could well see countries argue over the costs, repercussions, risks, liability, and control of such measures. Countries such as China, India, or Bangladesh largely depend on climate-sensitive economies, making them vulnerable to the impact of large-scale geoengineering projects. The mentioned Environmental Modification Convention was in fact the response of the United Nations to the U.S. Pentagon's project Popeye in the early 1970s, aiming at using cloud seeding techniques to increase the strength of the monsoon rains to disturb the Ho Chi Minh Trail.[65] This project went nowhere, but the risk of political conflict over the control and impact of geoengineering remains.

Moreover, any addition of sulfate aerosols to the atmosphere would whiten the sky, which would presumably make sunsets redder and more colorful. Walsh mentions that the famous painting *The Scream* by the Norwegian artist Edvard Munch was inspired by the bloodred afterglows in Scandinavia that scientists believed to be connected to the 1883 eruption of the volcano island of Mount Krakatau in Indonesia.[66] Although this may seem trivial in comparison to some other considerations, Robock argues that it "could have strong psychological impacts on humanity." It is a well-known fact that sunlight influences human behavior and mood. For example, scientists believe that one of the factors leading to Finland's high suicide rate is lack of sunlight, and Estonia, where a fourth of the population suffers from Seasonal Affective Disorder, holds an annual winter light festival in the capital of Tallinn to cheer up its residents.

NUCLEAR PROLIFERATION AND THE RISKS OF NUCLEAR ENERGY

While for many who grew up in the Cold War era, nuclear power is synonymous with disasters such as the Three Mile Island (USA 1979) and

Chernobyl (former Soviet Union 1986) nuclear accidents, nowadays attitudes seem to be changing within the environmentalist community as well as among policy makers regarding nuclear power. In 2006 the *Washington Post* published an article titled "Going Nuclear: A Green Makes the Case" by Patrick Moore, one of the co-founders of the environmental activist group Greenpeace. In this article, Moore explains that although he used to be opposed to nuclear energy, he has changed his mind in light of its role in mitigating climate change. He claims that "nuclear energy is the only large-scale, cost-effective energy source that can reduce these emissions while continuing to satisfy a growing demand for power. And these days it can do so safely."[67] Moore addresses a number of concerns regarding nuclear power, and concludes that nuclear power is affordable; that both nuclear plants and waste can be made secure; and that nuclear reactors are built to security standards that would not make them seriously vulnerable to a major terrorist attack. He goes so far as to claim that the 1979 Three Mile Island nuclear reactor core meltdown was "in fact a success story" in that there were no injuries or deaths.[68] Along these lines, well-known ecologist James Lovelock, who shaped green thinking, has openly argued that nuclear energy is the only feasible alternative to fossil fuels: "I'm absolutely sure that if we—by which I mean civilization—are going to get through this century without some greenhouse disaster, we've just got to use nuclear on a grand scale. Renewables simply can't do it in time."[69] The former United Nations weapons inspector in Iraq, Hans Blix, perceives climate change as more dangerous than nuclear weapons proliferation: "While I would be the last to underestimate the risk of proliferation of weapons of mass destruction, in particular nuclear weapons, I think the environmental risks we face are even greater."[70] However, these views are singular opinions and do not necessarily express public opinion, which is much more critical of the benefits and risks of nuclear energy.

There is indeed a consensus that nuclear fuel can and does provide zero-emissions energy at a relatively low cost, and has the capacity to supply significant amounts of energy—something that we cannot say for most other alternate energy sources. Nuclear power is also a technology that has existed and been in use for decades, and thus the technology is at an advanced stage. However, despite Moore's arguments, many believe that nuclear power is not entirely benign and point to serious security-related issues involving nuclear power.

Nuclear energy in many countries is a substantial contributor to the total electricity supply. In the United Kingdom, for instance, nuclear energy contributed 22 percent of the country's total electricity supply in 2004. This percentage reduced the country's emissions by between 7 and 14 percent. However, 11 of the United Kingdom's first-generation "Magnox" nuclear reactors will be phased out by 2010, others are scheduled to follow, and with only one-quarter of the total nuclear capacity remaining by 2020 there is an urgent need to develop new sources of electricity supply. Yet, a 1994 government paper that examined the economic and commercial viability of new reactors concluded that there is little government support for new nuclear power stations.[71] Other countries followed similar policies, such as the former green-social-democratic coalition in Germany, which passed a bill to phase out nuclear energy altogether. But given many industrialized countries' ambitious emissions reduction targets—the United Kingdom announced a 60 percent reduction by 2050—there are serious doubts that this can be achieved solely with increased energy efficiency, increased renewable energy sources, and increased use of gas-fired power stations.[72]

Besides the economic costs of start-up and legal regulatory and decommissioning costs, one of the greatest challenges of nuclear energy is the disposal of nuclear waste. So far, long-term storage of nuclear waste is very costly and bears radioactive spillage risks. Another concern is the risk of nuclear accidents as happened in Chernobyl in 1986. A public survey in the United Kingdom reveals that many interviewees associated radioactive waste with Chernobyl, indicating how deeply Chernobyl has entered the public psyche as a global disaster.[73] When asked about the perceived risks and benefits of nuclear energy as a contribution to a low-carbon society, the survey results demonstrate that radioactive waste is considered more negatively overall than climate change. Members of the survey's focus groups reacted to radioactive waste with intense dread and fear associated with catastrophe, death, and institutional failure connected to tangible historical incidents. By contrast, participants did not see climate change as an immediate threat and it lacked personal immediacy.[74]

Another far less sinister but equally frightening worry is that of earthquakes and other natural disasters causing damage to nuclear power plants. In July 2007, Japan was abuzz with concerns about the safety of the Kashiwazaki-Kariwa nuclear power plant in Niigata Prefecture after a devastating earthquake of 6.8 on the Richter scale hit the area. The London-based World Nuclear Association, an advocacy group backed by power supply companies in the United Kingdom, claimed, "All the functions of shutdown and cooling worked as designed. . . . While there were many incidents on site due to the earthquake, none threatened safety and the main reactor and turbine units were structurally unaffected."[75] Nevertheless, the *New York Times* reported that the plant "suffered widespread damage, including minor radiation leaks, ruptured pipes, flooding and a fire that belched black smoke for more than an hour on live television."[76] Thus, despite the absence of casualties, the damage caused by the earthquake and reports that the earthquake was "more than twice as strong as the plant's design limits"[77] caused an increased sense of insecurity in public opinion regarding the safety of nuclear power. Japan, of course, is one of the most developed nations in the world and one of the best equipped in terms of earthquake safety. But nuclear power is not limited to wealthy nations with such capabilities; thus there are concerns about what would happen if a similar incident occurred in a less developed country.

Perhaps the issue receiving the most international news coverage is that of "rogue" states and the ambiguity of their nuclear programs. While North Korea likes to parade its nuclear arsenal on the streets of Pyongyang to cheering crowds, proudly proclaim its capabilities to the world, and test its Intercontinental Ballistic Missile (ICBM) technology by firing rockets over Japanese airspace into the Pacific, other nations are far more subtle and clandestine about their nuclear programs.

Foreign Policy magazine ranks countries according to their stage of state failure in their "Failed States Index 2009." The Index ranks countries on a continuum and classifies them according to their level of state failure, from stable, borderline, in danger, and critical, the category closest to failure. China, India, and Russia—countries that possess nuclear power stations— are ranked "in danger," whereas Pakistan is the only country with nuclear facilities that is labeled "critical." Bangladesh, Belarus, Egypt, Indonesia, Iran, Israel (along with the West Bank), North Korea, Thailand, Vietnam, and Turkey are all countries that belong to the former category with planned or proposed nuclear reactors. This long list of countries, some of which are new aspirants and are politically unstable, to acquire nuclear facilities raises further concern in terms of nuclear safety. More importantly, this increases the risk of countries and terrorist groups acquiring nuclear material or waste and constructing a nuclear weapon with enormous destructive potential.

Iran, for example, is widely suspected by the United States and Europe of attempting to develop nuclear weapons, despite claims by the Iranian government that its nuclear program is peaceful and exclusively devoted to generating civilian nuclear power. Although the International Atomic Energy Agency (IAEA) has uncovered a number of instances in which Iran failed to declare certain activities, it has failed to uncover irrefutable evidence of noncivilian nuclear activity. The U.S. Central Intelligence Agency (CIA) has uncovered many activities it deems suspicious, including instructions from a Pakistani nuclear scientist for shaping uranium metal into "hemispherical forms"—a technique usually used to shield the core of a bomb.[78] The debate nevertheless rages on, with Iran categorically rejecting all accusations of noncivilian nuclear research. Many in the United States are far more skeptical of CIA suspicions of Iranian nuclear activity after the agency failed to find weapons of mass destruction in Saddam Hussein's Iraq. Regardless of Iran's real intentions, it is obvious that states' nuclear ambitions are often ambiguous.

An underlying issue regarding nuclear power is that because it is so efficient, relatively cost-effective, and arguably a tool in combating global warming, more and more states are bound to acquire such technology.

And because nobody has the right to tell one state that it is developed enough for nuclear power, and another that it is too "rogue" to be trusted, or does not have the infrastructure to make it safe enough, nuclear power-related security remains a concern.

CONCLUSION

The aim of this chapter was to examine the unintended consequences of climate change mitigation. Some of the most important mitigation strategies, including first-generation biofuels, geoengineering projects, and nuclear energy have ripple or "knock-on" effects with implications for human, national, and international security, as summarized in Table 5.1. Other mitigation strategies such as hydropower, energy efficiency, wind, and ocean energy are less of a concern for human or international security, and therefore were not included in the discussion.

It appears that the most problematic of all mitigation strategies is the use of nuclear energy and the risks associated with it. The challenge to dispose of radioactive waste and the risk of nuclear accidents are well-known and remembered by the public, posing the main obstacles to commissioning new nuclear reactors. Another fear is the spread of fissile material, of weapons-applicable nuclear technology and information, and the risk that terrorists may construct a so-called dirty bomb by combining radioactive material with conventional explosives. This risk increases when fragile states such as Pakistan, North Korea, or Iran employ nuclear facilities, both for commercial and military use, in the absence of a stable and predictable political regime. When states break down, as the former Soviet Union did, there is the danger of a power vacuum leading to the disappearance of radioactive waste, other nuclear material, or technology that could get into the wrong hands.

Carbon sequestration and geoengineering projects largely pose a risk for ecosystems we depend on. Less of an issue of international security, these mitigation strategies alter our planet's biochemical cycles on a large scale. Ocean fertilization and the injection of carbon into the deep ocean bear heavy environmental risks, and their effectiveness remains disputed. By contrast, carbon sequestration in geological formations appears to be safer. Solar radiation management schemes are relatively low-tech and affordable. To add sulfate aerosols to the atmosphere is technologically feasible. However, the environmental and political implications are enormous: Environmentally, such intervention can disturb other important hydrological cycles, such as the Indian Monsoon. In turn, the change or collapse of the Indian Monsoon could have severe economic and political consequences, especially when nations not affected by such change have been involved in such intervention. Non–carbon-related geoengineering projects that reflect sunlight, thereby interfering with the Earth's energy

Table 5.1 Unintended Ripple Effects of Climate Change Mitigation Strategies

Strategy	Type of Mitigation Strategy	Description of Strategy	Intended Effects	Ripple Effects
Biofuels	Renewable energy source	Solid, liquid, or gaseous fuels are generated from recently lifeless or living biomass.	Biofuels emit much less carbon dioxide than fossil fuels, and would reduce greenhouse gas emissions significantly if used as the dominant fuel.	N_2O from fertilizer, if released into atmosphere, would cancel out the cooling effect of fossil fuel savings.
		First-generation: generated from fermentation of crops with high sugar content, or from plants that contain vegetable oil. Oil can be used directly as diesel or refined to produce biodiesel.		Land used to grow biofuel competes with land usage for livestock, livestock feed crops, and human food crops, and contributes to high food prices. Food insecurity can lead to civil unrest, exacerbate preexisting tensions, and cause nations to adopt protectionist economic measures.
		Second-generation biofuels are produced from nonfood biomass such as wood, agricultural residuals, municipal waste, and microalgae.		Deforestation can occur to make way for biofuel crop plantations. Deforestation is a major cause of climate change.
				Some second-generation biofuels require pesticides and fertilizers.
				There is concern that removal of biomass from forests for second-generation biofuels may reduce biodiversity and decrease soil quality.

Nuclear Power	Alternate energy source—avoiding emissions	Nuclear power generates little to no greenhouse gas emissions and does not contribute to global warming.	There are concerns regarding safety of nuclear plants: leakage, explosions, vulnerability to terrorist attacks. There are also concerns relating to nuclear waste disposal and health hazards. A risk exists of nuclear power in the hands of rogue states and organizations, that could divert civilian nuclear power to make nuclear weapons. States such as Iran could use a civilian nuclear program as a cover for nuclear research for military purposes.
Ocean Fertilization	Carbon dioxide sequestration—reducing emissions	Iron, ammonia, or other chemicals are added to the ocean.	The "fertilizers" would increase phytoplankton biomass and productivity, which would speed up the biological pump that sequesters carbon in the deep sea.	This strategy could cause eutrophication of the seas, resulting in deep-sea hypoxia (oxygen depletion), which would shift the microbial community toward methane and nitrous oxide producing organisms. This would alter food webs and biogeochemical cycles in unintended ways.
Solar Radiation Management	Geoengineering—not related to emissions reduction or avoidance	Numerous strategies include:	This plan aims to decrease Earth's albedo, or reflectivity, thus counteracting the effects of global warming by decreasing the energy absorbed from the sun.	Sulfate aerosol loading would whiten skies and diminish direct solar beam, solar technologies much less effective.

(Continued)

Table 5.1 Unintended Ripple Effects of Climate Change Mitigation Strategies (Continued)

Strategy	Type of Mitigation Strategy	Description of Strategy	Intended Effects	Ripple Effects
		Globally load sulfate aerosol particles in the stratosphere.		Because less solar radiation reaches the surface, less energy is available for the hydrologic cycle.
		Soot, which would offset reduction in ozone, has been proposed as an alternative to sulfates.		Aerosols can exacerbate loss of ozone from stratosphere.
		Alternatively, load sulfates into the troposphere.		Tropospheric loading could cause adverse health effects, reduced visibility, and increased acid rain.
		Launch millions to trillions of large high-altitude balloons.		Balloons might interfere with communications.
		Inject particles into the mesosphere.		
		Cover ~70,000 km^2 of desert with large reflective Mylar® film.		Mylar® in desert would disrupt local weather patterns.
		Float a reflective structure in the ocean, the size of a continent.		Floating reflective structure would disrupt hydrologic cycle, weather, and ocean currents.

balance, are rather speculative than ready for commercial use. Of course, simple solutions exist, however less effective, such as painting roads and buildings white.

Biofuels can play a significant role in mitigating climate change as long as they do not compete with food crops or areas of biodiversity, such as pristine rainforests. Second-generation biofuels may be a solution, although they also require fertilizers and other inputs derived from fossil fuels.

So, what is the best way forward? To better access the effectiveness of mitigation strategies, pilot studies, such as the Indo-German Lohafex research project examining ocean fertilization to stimulate algae growth, will improve our knowledge about the efficiency of these technologies before commercial introduction. In addition, these studies help in assessing the associated environmental risks. We need to diversify our energy sources, using a mix including renewable and fossil fuels. Governments need to establish incentives and establish effective compliance mechanisms. Second-generation biofuels, carbon sequestration deep underground, and some of the localized solar radiation management technologies seem to be the most feasible, with few implications for human and international security. On the local level, there are already many initiatives under way. What is needed is help from policy makers to coordinate and to support these efforts, such as small-scale hydropower stations as seen in Norway or the use of biogas in China.

There is no "ideal" mitigation strategy; one always needs to weigh the benefits against the risks. In addition to the actual and perceived security risks of climate change mitigation strategies, other considerations matter: For many governments, the focus is on energy security. For example, the U.S. government expressed interest in becoming less dependent on oil from the Middle East. Other countries, largely in the developing world whose carbon footprint is low, such as the Philippines, prioritize biofuels as a strategy to boost rural development.

As said, there is no best way to avoid or reduce greenhouse gas emissions; each strategy will have its risks and benefits. Science can help us better assess the unintended consequences of our efforts to combat global warming.

CHAPTER SIX

The Way Forward: A New Environmental Security Agenda for the Twenty-First Century

Every nation on this planet is at risk. And just as no one nation is responsible for climate change, no one nation can address it alone.

President Barack Obama, July 9, 2009, L'Aquila, Italy

INTRODUCTION

Global implementation of innovative technological, legal, and international governance solutions is necessary if the human species is to respond and adapt successfully to climate change in the twenty-first century with a minimum of international and local conflict, violence, and economic loss. There is consensus that climate change is inevitable if we continue with a business-as-usual scenario. Although mitigation, the reduction of greenhouse gas emissions, is important, responding and adapting to climate change will be crucial to avoid adverse outcomes of climate change impacts.

One way to deal with the—largely—human security implications of climate change is the application of preventive measures. We have made huge strides in weather forecasting and climate modeling, as well as in research on socioeconomic indicators. Forecasting and early-warning systems can help to predict climate change impacts. This approach is important; it is proactive rather than reactive, ultimately helping to prevent food shortages, forced migration, loss of life, and even violent conflict arising from climate change impacts and additional stressors, such as poverty or poor health.

Early-warning systems can help reduce the impact of sudden meteorological hazards. In the long term, we need to adapt to change. In fact, climate change adaptation is already happening, and some developing and developed countries are implementing limited measures.[1] The following section outlines available adaptation strategies, ranging from financial to infrastructural and economical, assessing the costs and benefits and highlighting the limitations and barriers of successful adaptation. The most obvious barriers to adaptation are financial, ecological, and physical in nature—but cognitive, informational, social, and cultural factors also play an important role. Developing countries especially are constrained by a multitude of factors: Typically, their per capita incomes are low, and additional stressors, such as poor health, lower educational standards, and ineffective governance impede successful adaptation. Adaptation takes place on all levels: Individuals, groups, and institutions will need to adjust to current and future climate change impacts.

Adaptation and response measures to climate change impacts are often voluntary and depend on the adaptive capacity, available options, and limitations. According to Nelson, Adger, and Brown, adaptive capacity provides the "preconditions necessary to enable adaptation, including social and physical elements, and the ability to mobilize these elements."[2]

If policy makers, scientists, and the public are serious about avoiding "dangerous" human-induced climate change, the need exists for an institutional, legal mechanism to prevent conflict and human insecurity arising from climate change impacts. For small island states, sea-level rise constitutes a national security issue, threatening their very existence and state sovereignty. Because climate change is global in its origins, we need to think about global solutions. The United Nations, with its specialized agencies, has global legitimacy, albeit depending to some extent on the type of organization. For instance, the United Nations Framework Convention on Climate Change (UNFCCC) and its secretariat has the legal and technical authority to address issues of compliance to agreed reduction targets. Others suggest that the United Nations Security Council should take a more proactive role in dealing with the security implications of climate change.

In 2007 and under the leadership of the United Kingdom, the United Nations Security Council held its first session on climate change and security. Some advocate the United Nations Security Council as the appropriate lead agency to deal with the security aspects of climate change but there is some controversy. The permanent members of the Security Council are the world's biggest polluters, yet typically they are not the most adversely impacted by climate change. For instance Africa, only represented by a three-member elected bloc on the Security Council, is widely seen as the world's most vulnerable region to climate change impacts, yet the least able to adapt and—most worrisome—with the greatest potential for violence. Many have argued for years that the Security Council does not reflect the shifting and dynamic new World Order. China, India,

Brazil, Germany, and Japan are important geopolitical players without permanent seats in the Council. Moreover, the Secretary-General, as head of the United Nations, has very little political power but rather moral authority. Perhaps a promising way forward is the evolving G-20 heads of states summit that could formulate policies to combat climate change, as it comprises the biggest polluters. Yet, we are left with a dilemma: Those who are most vulnerable to climate change, the less developed regions, are currently not represented in this forum.

EARLY-WARNING SYSTEMS

Parallel to the discussion on adaptation, mitigation, compliance, and the establishment of new legal instruments, there are efforts to improve the prediction of changes so that societies and decision makers can respond to them. Advances in modern science and satellite imagery have improved meteorological forecasting. Clearly, it is easier to forecast the weather than social interactions, such as political upheaval triggered by food shortages. Still, connecting the power of meteorological forecasting to socioeconomic analysis could be a useful way forward, especially in countries with climate-sensitive economies.

One suggestion put forward by the Center for International Earth Science Information Network (CIESIN) at Columbia University's Earth Institute is to combine meteorological data with conflict data. This analysis could be one possible way to identify countries and regions at risk: Regions, in particular those economically dependent on agriculture, that receive very little precipitation (far below normal) are at greater risk to lose economic income due to drought. In sub-Saharan Africa, as illustrated in Figure 6.1, most regions are at a far greater risk to lose their economic income due to drought than compared to other natural hazards, such as tsunamis, earthquakes, volcanic eruptions, tropical cyclones, and floods. If those regions overlap with conflict zones, they become hotspots that require the special attention of the international community to prevent an escalation of the existing conflict. For instance in West Africa, Guinea and the Ivory Coast had both internal conflict in 2006 as well as below normal rainfall between August 2006 and January 2007, based on a 1961 to 1990 average. Several international agencies have already developed appropriate tools for weather-related warning systems.

For instance, the UN Environment Programme/Global Resource Information Database (GRID) has developed a Disaster Risk Index. The World Bank, in collaboration with CIESIN at Columbia University, has undertaken a Hotspots Analysis to better understand the impact of natural hazards on social interactions, including economic loss and mortality.[3] USAID uses its Famine Early-Warning system. There are also regional systems such as the Conflict Early Warning and Response Network

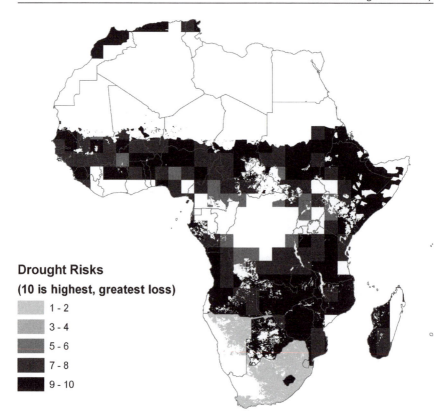

Figure 6.1 Economic Loss Risk in Africa. Compared to other parts of the world, Africans have a high risk of losing their income due to drought. Other types of natural disasters, such as tsunamis, earthquakes, floods, and windstorms, are less significant. Many African economies depend on agriculture, making them more vulnerable to climatic changes. The areas with the highest loss risk are also some of the most densely populated areas, such as Ethiopia or southern Nigeria.

Sources: Center for Hazards and Risk Research (CHRR), International Bank for Reconstruction and Development/The World Bank, Washington, DC, Center for International Earth Science Information Network (CIESIN), The Earth Institute, Columbia University, Palisades, NY.*

*Drought measure: 1980–2000, three-month average precipitation; a drought event was identified when the magnitude of a monthly precipitation deficit was less than or equal to 50 percent of its long-term median value for three or more consecutive months. No data result from a mask that excluded low population density and without significant agriculture. All grid cells are divided into 10 groups, called deciles. The data are better suited for relative comparisons than for absolute figures. Reporting actual GDP figures would portray an unrealistic impression of precision. The more modest objective here is to provide a relative representation of disaster risk. For cartographic output and interpretation, therefore, the authors convert the resulting numbers into index values from 1 to 10 that correspond to deciles of the distribution of place-specific aggregate GDP figures.

(CEWARN) for the Horn of Africa.[4] A report by the United Nations University Institute for Environment and Human Security (UNU-EHS), CARE International, and CIESIN maps the effects of climate change on human migration.[5] Other agencies, such as the Food and Agricultural Organisation (FAO) and the UN Office for the Coordination of Humanitarian Affairs (UNOCHA), maintain information networks that allow them to monitor agricultural developments and information on natural disasters and complex emergencies, respectively. All these efforts are important; what is lacking is a standardized tool easily accessed and used by policy makers.[6]

RESPONDING AND ADAPTING TO CLIMATE CHANGE

In addition to early-warning systems that improve forecasting of climate-related natural hazards such as floods or droughts, societies need to be prepared for environmental changes to avoid unwanted outcomes. For thousands of years, humans have adapted to environmental changes. Mobility was one of the main strategies during the ice ages to cope with the changing ecological situation. Other strategies involve crop diversification, irrigation, water management, disaster risk management, and a more contemporary option, insurance.[7] Some of the anticipated changes, such as heat waves, glacier retreat, melting of the Arctic ice sheets, and increased tropical cyclone intensity are novel in the sense that humankind has very little experience in dealing with their effects. Depending on economic strength and the availability of information and technology, societies can respond to climate change impacts and adapt to them.

On parallel, individuals, firms, and governments aim at reducing the output of greenhouse gas emissions. Energy efficiency measures and switching to renewable energy sources are examples. By drastically reducing greenhouse gas emissions, societies can influence the rate and magnitude of change through abatement of greenhouse gas emissions (mitigation), and prepare for changes and adapt to them (adaptation).[8] In this respect, climate change adaptation is complementary to mitigation.[9] Yet, adaptation receives less attention in the debate dealing with climate change impacts. The United Nations climate change regime, for instance, considers adaptation as a "second response option," mainly emphasizing mitigation strategies.[10] Most important, adaptation will be necessary and often the *only* response option for developing countries with climate-sensitive economies. In a developing country context, mitigation is less important; developing nations are typically countries with a low carbon footprint and limited abatement potential.

Despite all mitigation and adaptation efforts, climate change impacts cannot be avoided, argues the IPCC.[11] Greenhouse gas concentrations in the atmosphere have increased steadily since recording started. Even if we drastically reduce emissions now, we cannot avoid a warmer, more variable, and much less predictable climate. Moreover, for some regions and countries, even adapting to climate change is not an option. A one-meter sea-level rise will submerge 80 percent of the Marshall Islands. To make matters worse, the Marshall Islands' adaptive capacity is lower than, for instance, that of Australia.[12] To better illustrate this, a Marshall Islands woman earns on average 8 percent of her Australian counterpart.[13] The scale of the problem is also different: An Australian may be worried about coastal flooding and water shortages, but for Marshall Islanders, the question is not how to adapt but how to survive. Similarly, in Bangladesh a 45cm sea-level rise could result in a 10.9 percent territory loss, making up to 5.5 million people homeless.[14]

But what is adaptation? There are various definitions, borrowed from and based on the evolutionary biology literature. For instance, the 2009 *Oxford English Dictionary* defines adaptation as "a change by which an organism becomes better suited to its environment." If we apply this concept to climate change, adaptation is a change, either spontaneous or deliberate, by which an individual, society, or institution better adjusts to climate change impacts or effects. In the context of climate change and security, adaptation is a means to reduce human insecurity. This notion is also reflected in the IPCC, defining adaptation as "actual adjustments, or changes in decision environments, which might ultimately enhance resilience or reduce vulnerability to observed or expected changes in climate."[15]

As demonstrated previously, most of the security-related aspects of climate change (food insecurity, droughts, floods, tropical cyclones, sea-level rise, and subsequent migration) are predicted to occur in poor nations with climate-sensitive economies, largely in sub-Saharan Africa and Southeast Asia. Adapting to the impacts and adverse effects of climate change is therefore a matter of survival, as poor nations cannot wait for the developed world to agree on reduction targets of greenhouse gas emissions. More importantly, they have very little leverage to influence the shaping of industrialized countries' climate change mitigation policies.

Adaptation is often a short-term solution that can hardly address long-term changes, such as sea-level rise.[16] Of course, there is plenty of room for adapting before relocation becomes inevitable. The range of adaptation strategies is wide, including improved water storage and conservation techniques, erosion control, construction of seawalls and storm surge barriers, improved sanitation, and the diversification of energy sources.[17] The following section provides a snapshot of what is possible.[18]

Practical Examples of Climate Change Adaptation

Some adaptation projects are already under way: In 1998 in Nepal, for instance, authorities have partly drained the Tsho Rolpa glacial lake to prevent a sudden and uncontrolled lake outburst with subsequent flooding.[19] This is an example of a proactive approach to prevent irreversible damages to ecosystems, biodiversity, and people's livelihoods. Reactive approaches or "wait-and-see" attitudes are less successful because they are not implemented until the damage has already occurred. Some climate-related events, however, are difficult to anticipate, such as landslides or flash floods. They require reactive solutions, including emergency response, disaster recovery, and migration.[20] Here, improvements in climate monitoring and remote sensing have helped to monitor natural hazards and weather phenomena like La Niña and El Niño. Today, climatologists can forecast these events several months ahead.[21]

Other practical examples include snowmaking in the Alps to keep attracting ski tourists even in periods of low precipitation. This requires building large water reservoirs to provide the snow cannons with the necessary water. In Germany, the authorities agreed to cover part of the only glacier with white reflecting foil to prevent the ice from melting in the summer. This is, of course, very costly and likely to remain an exception.

There are also innovative financial instruments to respond to climate variability. One such measure is insurance for property, health, and crops. Insurance can help to spread the risk and avoid financial hardship. In fact, most adaptation strategies link to insurance. Insurers have developed risk assessment tools and helped to transfer adaptation incentives to individuals and companies. Re-insurance companies have increased their capacity for climate-related risk analysis.[22] In Africa, where a large percentage of the population depends on natural resources for their livelihoods, index-based weather insurance bundled to microcredit for agricultural inputs can help to offset droughts. In Malawi, researchers combined climatic forecasting, agricultural, and financial models to adjust the amount of high-yield agricultural inputs given to farmers. By doing so, wealth accumulation would increase, theoretically, therefore helping farmers to cope with adverse weather conditions and reducing their vulnerability.[23]

Other measures address health issues. Heat waves that occurred in 2003 in Europe and in 1998 in Toronto, Canada, have led to the development of hot-weather alert plans. These plans use early-warning systems and respond with public health adaptation measures, targeted especially at the elderly to protect them from the heat.[24]

Also, long-term infrastructure adjustments are under way. For example, in Canada, authorities increased the height of the Confederation Bridge by one meter to increase the vertical clearance for vessels accounting for a maximum of one meter sea-level rise.[25] Most of the infrastructure projects aim at adapting to sea-level rise by increasing the height of dams in low-lying coastal areas or adjusting the height of coastal highways and other transportation systems.

Some of the adaptation strategies, such as safe water and improved sanitation, are also part of other development strategies to foster economic development. Often, they fit into broader national planning, for instance in the Netherlands or in Norway where climate change scenarios have been included into national coastal defense plans.[26]

Limitations and Barriers to Climate Change Adaptation

Adaptation has its limits and barriers. The implementation of some of the above-mentioned strategies depends on the underlying policy framework of each country and the availability of funds, technology, and other informational considerations. These are some of the factors shaping the adaptive capacity of a country to respond. Limitations to adaptation are physical, ecological, financial, cognitive, and social in nature.

Physical and Ecological Limitations

The physical and ecological limits depend on the magnitude of climate change, and there are thresholds beyond which some species are unable to adapt. The same applies to physical infrastructure: Once the sea level rises above a certain threshold, roads and dams are no longer able to protect the land or secure transportation. Although it is technically possible to protect low-lying areas from a dramatic, possibly several meters sea-level rise, it is probably not economically feasible, so retreat remains the only option.[27] In the area of tourism, for instance, indoor skiing facilities have increased in numbers, even in desert regions like the emirate of Dubai. However, there is no question that such facilities cannot substitute for the loss of income of communities that depend on winter and ski tourism, and will remain an exception. Tropical, low-lying islands, such as the Seychelles or the Maldives, face a severe threat to tourism, although they often depend financially on this sector.

Financial Limitations

There are various reasons for financial limitations to climate adaptation: One is the national and global costs involved in climate change adaptation and impact research. Here, the critical question remains, how much impact is being avoided by adaptation? Whereas some adaptation measures are relatively inexpensive, others are prohibitively expensive; again, others are simply not feasible technologically even if unlimited funds were available. Initially, adaptation costs are lower than the cost of repairing damage, but once the adaptation costs equal the cost of damage, it is not clear how much individuals, firms, and governments are willing to pay for additional, feasible adaptation.[28]

Several studies tried to estimate global adaptation costs. The World Bank roughly estimates global costs to adapt investments in developing countries to climate change at between US$4 billion and US$37 billion per annum.[29] This figure only reflects investments and does not include the cost of additional impacts. Additional impact costs are much higher, ranging from US$10 billion to US$100 billion according to World Bank calculations.

Another estimate comes from UNFCCC's National Adaptation Programmes of Action (NAPA), which are designed for Least-Developed Countries (LDCs) to identify priority areas to respond to immediate and urgent climate change risks. As of 2009, five countries completed NAPAs, according to the Stern report.[30] Each country calculated adaptation costs of an average US$25 million. Extrapolating this figure, this would mean an adaptation cost of US$1.3 billion for all 50 LDCs.

In addition, the UNFCCC secretariat commissioned six studies to estimate the additional investment needs and financial flows needed by 2030 for climate change adaptation. The UN secretariat differentiated the costs according to sectors, including agriculture, water, human health, coastal zones, and infrastructure. According to this report, the global costs range from US$49 billion to US$171 billion with roughly equal shares for developed and developing countries.[31]

A recently published report argues that a caveat of this study is the omission of important climate-sensitive sectors, such as mining and manufacturing, energy, retail, the financial sector, tourism, and ecosystem protection.[32] Therefore, cost estimates are probably too low, since some sectors are missing and others underestimated. For instance, coastal infrastructure protection could be more expensive when coastal landscapes must be protected for recreational purposes. Stronger storms, too, can damage coastal ecosystems in addition to sea-level rise. Ecosystem protection—if included—would substantially add to the costs.[33] Although health was included in the UN study, it only included the effect on three health issues—diarrhea, malaria, and malnutrition in low- and middle-income countries. It did not estimate the health costs for high-income countries, for instance, the impact of heat waves or for other climate-related health issues.[34] Moreover, the UNFCCC estimate omits the cost for disaster preparedness and contingency planning. Disaster management is clearly important to the discussion of how we adapt to climate change. Although many blame the U.S. Army Corps of Engineers for the New Orleans disaster, was it not the general absence of effective disaster response and contingency planning and coordination that caused this catastrophe? Here, the financial investments in governance and technical capacity to respond and operate coastal defense infrastructure more effectively should be included in the adaptation costs.[35]

The wide range of the above-mentioned figures says much about the uncertainties in calculating the true adaptation costs. Undoubtedly, the burden on weak economies in the developing world is high, especially given the multitude of development challenges impeding successful adaptation. The lower UNFCCC estimate would mean a doubling of current Overseas Development Assistance (ODA). Considering this figure, a lower estimate hints at the type of barriers developing countries face to adapt to climate change to keep the damage to a minimum.

Accordingly, low-income countries have serious financial constraints to adapting to climate change impacts. For example in Africa, farmers do not have the means to afford climate-risk insurance or inputs to diversify agricultural production. Droughts are high-loss events but do not occur frequently. Farmers are reluctant to pay for premiums covering low-probability but high-loss events.[36] In addition, smallholder farmers often lack access to microfinance schemes to invest in adaptation measures such as increasing water-storage capacities. Irrigation systems can help to cope with climate variability, but they are capital intensive and require maintenance and a basic level of management. Educational barriers, especially the lack of higher education institutions, prevent individuals from acquiring skills that could be useful for other sectors of the economy, thereby diversifying the agricultural-based economy. Other financial barriers concern the health sector. Individuals often cannot afford the most basic and inexpensive adaptation measures, such as, for example, insecticide-treated bed nets that can prevent the spread of malaria.

Informational and Cognitive Limitations

There are also informational and cognitive limits to adaptation. The scale and speed of the physical impact of climate change is uncertain. Uncertainties make it difficult to calculate the costs and benefits of adaptation.[37] Even if some of the outcomes were highly likely and this information is made available, it does not necessarily lead to adaptation. The perceived risk of climate change is context specific.[38] What is more, individuals' societal values, personal preferences, and experience guide their interpretation of information and subsequent decisions.[39]

The IPCC *Fourth Assessment Report* lists four important informational and cognitive barriers to climate change:[40] First, even if individuals are aware of the danger of climate change impacts, they do not necessarily react to it. A partial explanation is that the nature of the changes is long-term and reaches beyond a human life span. Consequently, individuals are reluctant to take proactive steps. Second, perceptions of climate risks differ depending on the societal context. Other stress factors, such as poverty, violent conflict, or a high disease burden can overshadow and minimize the perceived risk that climatic changes pose.[41] Third, perceptions about climate change impacts and the capacity to adapt to them do matter.[42] Interestingly, there is evidence that those who perceived themselves vulnerable to climate change often see themselves as victims of injustice, and at greater risk from other environmental hazards.[43]

More specifically, studies reveal that perceived risk is closely associated with race and gender. In the United States, most nonwhites and females perceive themselves as being exposed to a higher environmental health risk. This research examines the "linked possibility that this demographic pattern is driven not simply by the social advantages or disadvantages embodied in race or gender, but by the subjective experience of vulnerability and by sociopolitical evaluations pertaining to environmental injustice."[44] The study reveals that white males see themselves as having a low exposure to environmental health problems, such as toxins, nuclear hazards, and coal and oil power plants compared to all other groups. By contrast, groups with strong feelings of discrimination and vulnerability relate closely to high perceptions of environmental health risks. In a nutshell, environmental risk perceptions depend on race, gender, vulnerability, and environmental justice. The difference in females compared to males in perceiving risk can be explained by their role as care providers and being responsible for their offspring. Consequently, differential risk perceptions can limit adaptive actions, as shown in Germany and Zimbabwe in relation to flood risks. Researchers found there that action was influenced by the perceived ability and actual (or observed) capacity to adapt.[45]

Fourth, climate change media coverage often appeals to fear and guilt. However, such reporting does not necessarily trigger the appropriate action, adaptation. In sum, adaptation is shaped by individuals' experience, risk perception, personal preferences, and motivation. After all, behavioral changes depend on many factors that vary among individuals and groups and are difficult to predict.

Social and Cultural Barriers

The social and cultural context is also important, argues the IPCC. It states, "individuals and groups may have different risk tolerances as well as different preferences about adaptation measures, depending on their worldviews, values, and beliefs."[46] Differences in access to decision-making processes and political exclusion constrain adaptive responses. Migration, for instance, is one of the responses to climate change but there are cultural and social implications of such migration. For example, in small island states, people may have to relocate to other parts of the region, or possibly to other countries, when sea-level rise reaches a certain threshold that would make continued settlement impossible. Resettling in another country, as proposed by New Zealand for the relocation of the people of Tuvalu, poses a completely new set of social and cultural challenges, such as integrating migrants into a foreign economy and society. In other cases, resource-dependent communities who live in climate sensitive environments often do not perceive climate change as an imminent risk to their livelihoods. In the northern Canadian forests, communities associate climate change with environmentalists, whom these communities see as opposition forces.

Vulnerability and Adaptive Capacity

Although some societies are highly adaptive to their environment, they become vulnerable through processes of economic globalization and other modernization processes.[47] Other factors, such as conflict and poverty, also change the adaptive capacity of communities. A good example is Somalia, where a long-lasting civil war has destroyed much of the adaptive capacity of the Somali pastoral and smallholder farming communities. Irrigation systems have been destroyed or are poorly maintained. Lack of funding and community organization prevents a restoration of the irrigation systems in southern Somalia. Armed militias often control floodgates and collect "taxes" for one of the numerous Somali warlords.[48] In the past, Somali pastoralists and farmers developed a sophisticated system of reciprocity and resource sharing systems that have broken down in the course of the civil war. The rule of force replaced customary law that once regulated access to water and grazing areas. Consequently, this has changed and weakened the adaptive capacity of Somalis to respond to climatic shocks, such as floods and droughts.

Another example is Haiti, where a combination of political instability over the past decades coupled with environmental degradation, mainly deforestation, and frequent natural hazards have led to a limited adaptive capacity. From the early 1980s, the accumulative gross domestic product (GDP) of Haiti has stagnated, and even dropped following the political turmoil in 1991 (see Figure 3.4). Other political, economic, and environmental stresses in Haiti exacerbate vulnerability to natural hazards. The political factors include weak governance, the destabilizing influence of outsiders, extra-legal criminal networks with vested interests, and the role of armed forces, including the UN, the military, and the police. Economically, the lack of public goods and community organization, high unemployment rate, economic inequality, and unfavorable terms of trade are all limiting factors. Environmentally, current natural disasters, environmental degradation dating back to the colonial era, unfavorable topography, and massive deforestation are all challenges for Haiti. Other countries with high vulnerability to climate change but low adaptive capacity are Bangladesh, Vietnam, the Philippines, and Indonesia.[49]

In a developing country context, poverty, a high disease burden, political instability, poor infrastructure, and current natural hazards are all factors that limit the adaptive capacity of a country to respond to climate

change. Moreover, developing countries often have a large sector of the
economy that is climate-sensitive. In consequence, they are at greater risk
to lose economic income due to natural disasters, events that are likely to
increase in intensity in the future. Improvements in the health and educational system can help to reduce dependence on agriculture. But not all
development efforts reduce vulnerability to climate change. For instance
in Borneo, Bangladesh, and Fiji, developers have cut down mangrove forests to make room for new settlements, hotels, and industry. The rapid
destruction of mangrove forests has destroyed much of the natural protection against floods, storm surge, and erosion. In other places, commercial
shrimp farming has destroyed fragile ecosystems that can act as a buffer
between the coast and human settlements.[50]

Climate Change Adaptation—An Opportunity for Action

Despite the limitations and barriers to adapting to climate change
effects, there are several successful adaptation examples in development, as
mentioned earlier. For instance in China, engineers have put in place a
cooling system to prevent the permafrost that is supporting a 550-kilometer
railway across the Tibetan plateau from melting.[51] Nepal is another example, where glacier lake outbursts threaten the country's hydropower stations. Since the destruction of the almost-complete Namche Small Hydro
Project, caused by the breach of a natural dam in a glacier lake, authorities
have started to plan hydropower facilities at low-risk locations. In addition, early-warning systems can allow time to provide extra security, partly
drain glacier lakes to avoid flash floods, and take flood control measures
downstream.[52]

For many societies, adaptation is not an option but a necessity. It is important to note that individuals, communities, and countries that are hard-
est hit by predicted climate change impacts have contributed marginally
to the problem and are least able to afford adaptation. In addition, as the
Stern report puts it rightly, "they can afford even less *not* to adapt."[53]

The adaptive capacity of a country, community, or individual is linked
to social and economic development, including resource base, social capital, institutions, type of governance, national income, health, and technology. Additional barriers and limitations relate to individuals' perceptions
of climate change risks, the availability of information, and personal preferences, in addition to social and cultural dimensions. There are currently
adaptation projects under way, both in developing and developed countries. They are small in scale; most of them are financial, but some are
technological or infrastructure projects. At this stage, it is too early to
monitor and evaluate their effectiveness in reducing social vulnerability,
but they are encouraging examples of proactive behavior. Moreover, the
costs and benefits of adaptation are poorly understood, especially in areas

that are not captured by the formal market, including ecosystem services or public health improvements.[54] Most important, uncertainties in observed changes complicate decision-making that would create a legal environment that facilitates adaptation and supports individuals and firms to cope with the costs.

THE UNITED NATIONS SECURITY COUNCIL DEBATE

Apart from adaptation and early-warning systems, there will be climate-induced security implications and ripple effects of climate change mitigation that need to be addressed globally. The possibility of thousands of people migrating within and across countries presents an enormous humanitarian challenge, especially forced migration in the aftermath of a major tropical cyclone or a severe drought. But large-scale loading of aerosols into the atmosphere or nuclear proliferation can constitute a "threat to the peace" and international security, a term that is contained in Article 39 of the United Nations Charter.[55] Chapter 7, including Article 39, mandates the United Nations Security Council to take all measures, including the use of force, to restore international peace and security.

In April 2007, the UN Security Council held its first open debate on climate change and international security. The Chair of this Forum, former British Foreign Secretary Margaret Beckett, warned the Council of "migration on an unprecedented scale because of flooding, disease and famine."[56] This could lead to intensified competition over food, water, and pasture. With 55 delegations attending the debate, it set a record in terms of participation, demonstrating strong interest from the international community on this topic.

The reason for the UK-led initiative is obvious: For years, the United Kingdom has been vocal in addressing and pushing the climate change agenda in high-level meetings. The country complies with its reduction targets under the Kyoto Protocol and set climate change high on the agenda during the 2005 G8 summit in Gleneagles, Scotland. For the Security Council debate, the United Kingdom urged members to focus on a number of drivers of conflict, including border disputes, migration, energy security, resource scarcity, societal stress, and humanitarian crises.[57] For the UK government, the main aim of this debate was to raise awareness of the climate change challenge.

During the debate, climate change was mentioned as a threat to international security, by UN Secretary-General Ban Ki-moon and the Papua New Guinea UN Ambassador Robert G. Aisi. The UN Secretary-General said that "projected climate changes could not only have serious environmental, social and economic implications, but implications for peace and security."[58] The Secretary-General fears climate change as an additional stress factor for "vulnerable regions that face multiple stresses at the same time—pre-existing conflict, poverty and unequal access to resources, weak institutions, food insecurity and incidence of diseases such as HIV/AIDS."[59]

The Forum would "expect the 15-nation body to keep the issue of climate change under continuous review, to ensure that all countries contributed to solving the problem and that those efforts were commensurate with their resources and capacities." It also expected the Council to "review particularly sensitive issues, such as implications for sovereignty and international legal rights from the loss of land, resources and people."[60] The expectation of small island states like Papua New Guinea, and other members of the Forum is based on the UN body's strength to take action, to sanction countries, or even to use force, a characteristic that the UN climate regime lacks. On the other hand, the Security Council does not have the technical expertise to deal with climate change and countries that do not comply with agreed emissions targets. In terms of legitimacy, most speakers agreed that the Security Council's involvement is consistent with the Council's current mandate.[61]

For several reasons, it is difficult to request the Security Council to address the security aspects of climate change. One of the reasons is the power representation in the Council. The five permanent members, who have veto rights—China, Russia, the United States, the United Kingdom, and France—are among the largest emitters of greenhouse gas emissions. Poor countries, most affected by the adverse consequences of climate change but with a low carbon footprint, have a limited influence on decisions taken by the Security Council.

Another concern is to identify the culprit of climate change. This is much easier when war crimes are committed. But anthropogenic climate change is less tangible, and global in its origins. One way forward is to link actions of the Security Council to the compliance mechanism of the (post-) Kyoto regime. If a country fails to comply, the Security Council could respond appropriately.

There are two major arguments as to why the Security Council should get involved in climate-related issues.[62] First, governments, such as the United Kingdom, argue that the aim of such a debate is to raise global awareness. Second, some governments want to deal with climate change in the framework of conflict prevention. Mitigating climate change, they argue, could prevent conflict and hence, contribute to peace and security. The countries favoring such an approach are Germany, France, and some developing nations, for instance Bangladesh, that are particularly affected by climate change. They also define security in a broad sense to include human security issues, such as food security or health concerns.[63] Small island states, facing loss of territory and population and considering climate change an imminent threat to their national security, largely favor the second approach. Consequently, they have a strong interest in request-ing the Security Council to take appropriate action rather than merely debating the problems.

Again, other countries from the developing world, such as India and China, which are typically large emitters, are opposing such calls as they see their economic development agendas and energy security policies jeop-ardized. Apart from the small island states, the opinions were clearly split between the North and the South. Seventy percent of all Northern speak-ers supported the Security Council taking a more proactive role, whereas only 29 percent of the speakers from the South, most from small island states, voted in favor of mandating the Security Council to take action.[64]

Although there is no international body that could take action against countries that are emitting more than the agreed-upon quotas, several other international organizations could deal with the security aspects of climate change. The most important is the UNFCCC and its Kyoto Protocol with its secretariat in Bonn, Germany. Other international agencies, such as the United Nations Environment Programme, the United Nations Development Programme, or the World Bank, could also play a constructive role.

Within the UNFCCC, there is room to develop new protocols that could deal, for instance, with climate refugees. This could be in the form of a separate, independent legal and political regime created under such protocol on the recognition, protection, and resettlement of climate refu-gees to the UNFCCC, argues Frank Biermann at the Free University of Amsterdam.[65]

Yet, such a protocol raises many questions, such as funding. Resettling thousands of people, possibly millions, who in the case of sea-level rise cannot return home, will require substantial funds. Innovative taxes such as an international air travel levy could finance such projects.[66] Already, industrialized countries have agreed to assist the most vulnerable countries to meet adaptation costs under the UN climate change convention. Hans Joachim Schellnhuber, director of the Potsdam Institute for Climate Impact Research (PIK), proposes a more practical approach, arguing at the Climate Congress in Copenhagen in 2009 that payments should be made according to accumulated national emissions. Hence, the United States should accept the majority of climate migrants.

Most important, a new protocol should not duplicate efforts of existing agencies. For instance, UN Development Programme and the World Bank, who have a strong operational mandate, could serve as implementing agencies to assist climate refugees. The United Nations High Commis-sioner for Refugees (UNHCR), the International Organization for Migra-tion, and the UN Environment Programme could provide technical and legal support.

Extending the rather narrow definition of refugees of the 1951 Geneva Convention Relating to the Status of Refugees would be a viable alternative to any additional protocol to the climate change convention. This, of course, is not without controversy. There is the risk that legally acknowledging climate refugees could cause governments to use policies to block incoming climate refugees and deny them access to welfare.[67]

In summary, climate change is no longer an issue of development or environment but is now one of international peace and security. Here, a broadened concept of security can capture much of the security-related issues of climate change, including food security, water scarcity, and migration. Most important, the UN Security Council open debate shows the need for an effective compliance regime. Although the Kyoto Protocol has such a mechanism, the so-called Kyoto Compliance Committee, it has not yet proven very effective. Unfortunately, environmental agreements are known for their low compliance records.[68] Still, establishing additional protocols within the UNFCCC to address other outcomes of climate change, such as migration, could be another step toward a new international environmental regime to deal with climate change.

Most importantly, the entering of climate change into the arena of high politics can help to raise awareness and increase the pressure on both developed and developing countries to work together toward a global climate deal. As former British Foreign Secretary Margaret Beckett rightly observed: "Climate change can bring us together, if we have the wisdom to prevent it from driving us apart."[69]

SYNTHESIS AND CONCLUSION

Climate change impacts are primarily a human security issue. Although some climate change impacts will have broader security implications, such as resource scarcity in resource-dependent communities coupled with weak governance structures, low incomes, and a history of conflict or large-scale cross-border migration, it is important to focus on individuals who are both agents of change and are affected by climate change impacts. This approach highlights the importance of climate change impacts on social interactions, rather than focusing on the causes of climate change. The aim of the analysis is not to "securitize" nontraditional threats such as human rights, transnational crime, and the environment to gain the attention of policy makers and to mobilize resources, but to better understand the implications of climate change for all human beings (with less emphasis on the state). Rather than considering climate change effects as threats to international security, there is a need to better understand the interactions between climate change impacts and human well-being. This shifts the focus to climate change adaptation and resilience, two concepts that are important in order to examine people's and ecosystems' vulnerabilities.

The previous chapters discussed the security implications of three distinct climate change impacts: resource scarcity, natural disasters, and migration. In all three areas, people's livelihoods and their well-being are affected adversely if the level of greenhouse gas emissions in the atmosphere is not stabilized and brought to a level that would avoid "dangerous" anthropogenic climate change. In addition, some of the mitigation strategies to reduce carbon dioxide (CO_2) emissions have ripple effects with serious implications for human security. Examples include the use of first-generation biofuels, large-scale geoengineering projects, or the proliferation of nuclear energy (see Figure 6.2).

Human beings have altered their environment for centuries. Even thousands of years ago, our ancestors, equipped with spears and stone axes, were able to transform natural habitats into cultural landscapes. In some cases, these changes led to the collapse of societies, as scholars believe to have happened on Easter Island.[70] It has been argued, first by Brander and Taylor in 1998, that humans overexploited the island's natural resources, leading to a dramatic population decline.[71] A once-thriving civilization found on Easter Island thousands of years ago dwindled and almost

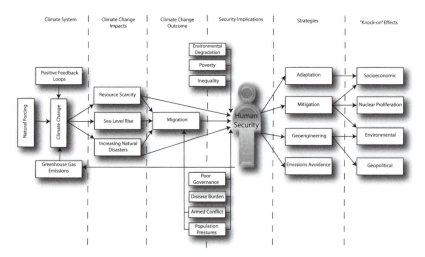

Figure 6.2 Human Security **Flow Diagram.** The possible interactions between the climate system, climate change impacts, migration, and human security illustrate how humans beings are affected by climate change as well as being agents of change. Yet, climate change solutions also have unintended consequences. The diagram summarizes the main issues pertinent to climate change and security implications.

Source: Adapted from Halvard Buhaug, Nils Petter Gleditsch, and Ole Magnus Theisen, "Implications of Climate Change for Armed Conflict," in *The Social Dimensions of Climate Change: Equity and Vulnerability in a Warming World,* ed. Robin Mearns and Andrew Norton (Washington, DC: World Bank Publications, 2009), 82."

disappeared when Europeans first stepped on land in the eighteenth century. In the Mediterranean region, vast areas became deforested long before the industrial revolution. At the same time, naturally occurring climatic changes have shaped societal change. Historical studies have shown the impact of climate change on warfare in Eastern China over the last millennium.[72] Earlier in time, research demonstrates how cooling periods can be associated with a higher frequency of armed conflict.[73]

What is new, is the human factor altering the entire earth system, not just shaping and changing landscapes and ecosystems—as happened for centuries—but altering global environmental processes. Most importantly, we are responsible for the increasing stock in greenhouse gas concentrations in the atmosphere, to an extent that these changes induced measurable impacts. The most notable are global mean temperature increase and sea-level rise; others include changes in the hydrological cycle, biodiversity loss, bleaching of coral reefs, more extreme temperatures, the loss of the Arctic summer sea ice, and the melting of the continental glaciers.

When It Is Too Late: Tipping Points and Security

What is even more worrisome is the emerging knowledge on tipping points, induced by negative feedback loops. Professor Schellnhuber of the Potsdam Institute for Climate Impact Research has been vocal in advancing this debate.[74] One possible scenario, he explains, is the dieback of the Amazon rainforest. Declining precipitation, a warming climate connected to the El Niño/Southern Oscillation and the conversion of forest into farmland, could bring the Amazon rainforest to a critical threshold. The loss of the Amazon rainforest would influence the global climate system and would result in the loss of biodiversity on an unprecedented scale. The best-known—and probably most advanced—tipping point is the loss of the Arctic sea ice. With a warmer climate, the summer sea ice is melting, exposing more seawater to sunlight. The dark patches of seawater absorb more solar radiation, hence amplifying the warming effect of the oceans. Over the last 30 years, the summer sea ice cover has decreased significantly, and in 2007 there was so little ice cover that the Northwest Passage was ice-free in September to October 2007.[75] Another observation is the melting of the Greenland ice sheet. Studies suggest that the melting of water generates a lubrication effect that accelerates the collapse of the Greenland ice sheet. A total collapse could lead sea levels to rise as much as seven meters.[76]

Science has linked these changes to human behavior, largely the burning of fossil fuels, deforestation, and agriculture. With this knowledge at hand, we were able to locate the source of the problem and could act accordingly. This understanding led to the formation of a global regime to combat climate change, the UNFCCC and its Kyoto Protocol. The ratification of the Kyoto Protocol entered into force on February 16, 2005. In Copenhagen in December 2009, the Conference of the Parties to the UNFCCC was supposed to negotiate a new international response to climate change, adopting a second-phase Kyoto Protocol, but did not reach any binding conclusion due to the controversy between industrialized, developing, and small island states. In parallel, emissions continued to rise and little has been done to reduce greenhouse gas emissions worldwide. Today, we are far from the Convention's objective, imposed in the second article, calling for a "stabilization of greenhouse gas concentrations in the atmosphere at a level that would prevent *dangerous* [emphasis added] anthropogenic interference with the climate system."[77]

How do all the above processes relate to security? As the United Nations Convention states, human interference with the climate system can be "dangerous," posing a threat to security. However, it is not clear what "dangerous" means and what levels of greenhouse gas emissions in the atmosphere constitute a threat to security. *Our Common Future* marked the beginning of officially introducing the term *environmental security* in 1987. This was largely due to the end of the Cold War, when conventional explanations of security became less relevant. Policy makers expanded the definition of security to incorporate environmental, demographic, and global health issues. When anthropogenic tipping elements of the earth system as described above materialize, these changes will most certainly undermine human livelihoods and state wealth, and expose humans to new disease vectors, constituting a threat and compromising security. It is less clear, however, how these changes invoke armed conflict; but there is evidence as demonstrated in this book that drastic rainfall drops can lower economic incomes in countries with climate-sensitive economies bearing serious political ramifications.

Making Hard Choices While Avoiding Unintended Consequences

In addition to the immediate threat of climate change impacts, there are serious security concerns of the discussed "knock-on" effects of climate change mitigation. Technologies to manage solar radiation are readily available at a relatively low cost. Altering the Earth's energy balance could influence global weather systems, such as the Indian Monsoon. If China decides to balance its CO_2 emissions by loading sulfate particles into the atmosphere, this could have climatic repercussions in other parts of world, triggering geopolitical tensions over "who's controlling the thermostat." Even storing CO_2 deep underground—a technology regarded as relatively safe—has come under criticism by studies that show that the injection of CO_2 can trigger earthquakes. First-generation biofuels have been heavily criticized for their limited CO_2 reduction potential and implications for food security. Moreover, the generation of biofuels threatens existing ecosystems, largely pristine tropical rainforests. So-called second-generation biofuels that use whole plants, microorganisms, or plant residues to generate energy are a more promising way forward. Finally, given the historical evidence of nuclear accidents and geopolitical considerations, the possibility of nuclear proliferation and nuclear energy to substitute for fossil fuels bears serious security implications, both for human and international security.

We are left with a dilemma. Clearly, climate change is a threat to human security, but many of the solutions themselves have security ramifications, as illustrated in Figure 6.2. Therefore, what is the appropriate way to avoid "dangerous" levels of greenhouse gas emissions? In the industrial countries, a drastic reduction of emissions is inevitable. To limit emissions, "cap and trade" systems as proposed by the U.S. Climate Security Act in 2007 and 2008 can create incentives for firms who are able to cut emissions.[78] But ultimately, we need to move towards a zero-carbon society. This can be achieved by employing a mix of proven renewable energies with little security risk, for example, using biogas when the wind speeds are too low to allow wind turbines to operate. Surplus electricity, for instance, at night could be used to pump water into reservoirs, storing electricity in forms of kinetic energy. Since a large proportion of emissions come from the energy sector, there is need to reform dramatically the way we move goods and people. Some solutions are the electric car, more mass transportation, and less air traffic. In the past years, the application of new energy sources for transport, such as hydrogen gas or fuel cells, has seen great advances in technological feasibility and efficiency.

Emissions also result from land-use change. For instance, a worrisome positive feedback effect is the potential release of large amounts of methane from melting permafrost and polar ice caps. While we can do little about these changes, we can accomplish more in other areas, for instance in the area of CO_2 sinks, especially in rainforests in tropical countries. The deforestation rate of tropical forests is proceeding at an alarming rate. To give an example, research shows that between 1950 and 2005, Borneo, an island shared by Malaysia, Indonesia, and Brunei, lost large expanses of tropical lowland and highland forests, including vast areas of rainforests.[79] On a regional level, several of the tropical countries in Africa and Southeast Asia lost forest cover in the past decade.

The UNFCCC is now considering a new mechanism to deal with emissions from deforestation and forest degradation of tropical forests in developing countries, commonly referred to as REDD (Reducing Emissions from Deforestation and Forest Degradation). It is estimated that in 2004 alone, emissions from deforestation were responsible for 17.4 percent of total greenhouse gas emissions, thus making deforestation and forest degradation the second largest source of greenhouse gas emissions after fossil fuels, namely oil, gas, and coal.[80] What is more, increasing demand for first-generation biofuels could further increase greenhouse gas emissions when developers clear tropical forests for palm oil plantations, agriculture, and other land uses. The purpose of REDD is to reverse this trend, by helping countries preserve tropical forests and generate forest carbon offsets, which may have the potential to be included in global carbon markets.

REDD could allow countries and individual companies to offset carbon credits against their emissions targets. There are additional environmental benefits, such as prevention of biodiversity loss, new opportunities for sustainable forest management, and eco-tourism. Successful REDD project implementation thus depends on the inclusion of indigenous and local communities who depend on forests for their livelihoods. REDD projects are already under way in Bolivia, Indonesia, and Australia. An Indonesian project covers a 770,000-hectare area of the Ulu Masen forest in Sumatra's Aceh province, home to endangered species such as orangutans, tigers, and elephants. To implement these projects successfully, local communities need to benefit from the preservation projects.

One potential risk for the successful implementation of REDD projects could be governance failure to ensure that forests are conserved. Another challenge is the engagement of local forest-dependent communities in a meaningful way to allow for benefit sharing and access while promoting the alleviation of rural poverty. There is fundamental criticism often voiced about the logic of REDD projects. REDD rests on the assumption of the "avoided bad" rather than the "committed good" of mitigation activities, such as new technologies, for instance, solar or hydrogen fuel. Besides REDD, many other activities could claim emissions avoidance, even measures that aim at reducing population growth in developing countries.[81] Another risk of REDD implementation is commonly referred to as "leakage." If forests are protected under REDD, deforestation and forest degradation may appear elsewhere. Also, it is not clear how long carbon will be stored in biomass. There is the likelihood that carbon stored in trees is released at a later stage, a phenomenon that experts refer to as "permanence."[82] Despite the technological challenges which appear now to have been addressed, and the operational and policy challenges, REDD presents a unique opportunity to address climate change, biodiversity loss, and water and soil conservation.

THE WAY FORWARD

Even if we radically change our behavior and thinking about mobility and energy use, we probably cannot avoid an increasing stock of CO_2 in the atmosphere. We may need to employ geoengineering technologies, some of which can be used locally and with known consequences. To successfully address climate change, individuals, firms, and policy makers need to work on local, regional, and international solutions. Many options, such as increasing energy efficiency or the use of hydropower, are proven technologies with known and often insignificant ripple effects. The application of such a mix of proven and inexpensive adaptation and mitigation measures in combination with an effective compliance regime can ultimately secure our planet's resource base for future generations.

Though technological solutions (with known consequences) to adapt to and to mitigate climate change exist, we are confronted with political, behavioral, and cultural barriers. The political challenges especially are difficult to overcome, as the ongoing discussions on the 2009 UN climate talks in Copenhagen have shown. Large emitters, such as the United States and France, demand that developing countries, such as China and India, agree to "meaningful" participation in the efforts to combat climate change. Who is going to share the enormous costs of climate change mitigation and adaptation? Developing countries rightly demand that rich countries help them finance clean technologies. In this context, it is instrumental to point to the security implications of climate change, shifting the debate away from technology and feasibility toward risks, vulnerabilities, equity, and justice.

Using the term *security* in the climate change discourse is not new. Detraz and Betsill ask the right question, whether "securitizing" the climate change debate has helped to solve the climate change challenge: "Does linking climate change and security concerns advance international efforts to address climate change? Is it possible to break through the political stalemate that characterizes the multilateral treaty-making process by raising the threat of climate change to the realm of high politics?"[83] Certainly, the UN Security Council debate on climate change and security raised awareness and lifted the topic into "high politics." But it remains open whether this discussion was an isolated event without the future involvement of the Council. The debate on climate change and security, however, remains important, as it emphasizes how human beings, in particular the vulnerable, are affected by issues such as food security, human health, and water stress. We need this debate; it is unlikely that climate change challenges can be met with conventional military operations. The authority for solutions is not to be found in the state military but rather in civil society and governments' bureaucracies.[84] Peter Liotta agrees: "The problems are too broadly distributed and the consequences are too deeply penetrating for such an approach to be successful."[85]

Liotta's statement also touches on the main critique of framing climate change outcomes in terms of security or human security. It has been argued that the concept is so broad that it becomes unusable.[86] Given the global implications of climate change, including resource scarcity, more intense natural disasters and sea-level rise as outlined earlier, affecting millions of people, it becomes clear that climate change will function as threat multiplier and in some cases invoke mortal struggles.

On the political level, the UN Security Council initiative lifted climate change into "high politics." There is clearly the need for a global social contract to combat climate change. The climate talks in Copenhagen in 2009—despite their shortcomings—were another attempt in the right direction, calling for a broader political commitment reflecting today's political realities. Governments are important, but even more, we need to focus on the people who are both agents and victims of climate change. Consumer preferences, voting behavior, and education are powerful tools in shaping our common future. Today's political leaders have come to realize that global problems require global solutions and that a handful of the most powerful countries in the world have failed—until now—to adequately address them.

NOTES

INTRODUCTION

1. Marc A. Levy, "Is the Environment a National Security Issue?" *International Security* 20, no. 2 (1995).

2. United Nations Development Programme, *Human Development Report, 1994* (New York: Oxford University Press, 1994), 22.

3. Roland Paris, "Human Security: Paradigm Shift or Hot Air?" *International Security* 26, no. 2 (2001): 88.

4. Nicole Detraz and Michele M. Betsill, "Climate Change and Environmental Security: For Whom the Discourse Shifts," *International Studies Perspectives* 10, no. 3 (2009).

5. Helge Brunborg and Henrik Urdal, "The Demography of Conflict and Violence: An Introduction," *Journal of Peace Research* 42, no. 4 (2005): 372.

6. Thomas Fingar. *National Intelligence Assessment on the National Security Implications of Global Climate Change to 2030.* Washington, DC: National Intelligence Council, 2008; National Intelligence Council, *Global Trends 2025: A Transformed World; NIC 2008-003* (Washington, DC, 2008).

7. Fingar, *National Intelligence Assessment.*

8. National Intelligence Council, *Global Trends 2025: A Transformed World; NIC 2008-003*, 66.

9. Geoffrey D. Dabelko, "Planning for Climate Change: The Security Community's Precautionary Principle," *Climatic Change* 96, no. 1-2 (2009): 15.

10. Daniel Deudney, "The Case Against Linking Environmental Degradation and National Security," *Millennium* 19, no. 3 (1990).

11. Koko Warner and Charles Ehrhart, Alexander de Sherbinin, Susana Adamo, and Tricia Chai-Onn, *In Search of Shelter: Mapping the Effects of Climate Change on Human Migration and Displacement* (Bonn, Atlanta, New York: United Nations University Institute for Environment and Human Security, CARE International, Center for International Earth Science Information Network at the Earth Institute of Columbia University, 2009): v.

12. Ban Ki-moon, "A Climate Culprit in Darfur," *The Washington Post*, June 16, 2007.

13. Alexander Carius, Dennis Tänzler, and Achim Maas, *Climate Change and Security: Challenges for German Development Cooperation*. Eschborn: Deutsche Gesellschaft für Technische Zusammenarbeit (GTZ) GmbH, 2008: 13.

14. Levy, "Is the Environment a National Security Issue?," 37.

15. Simon Dalby, *Security and Environmental Change* (Cambridge, UK; Malden, MA: Polity, 2009).

16. Jared M. Diamond, *Collapse: How Societies Choose to Fail or Succeed* (New York: Penguin, 2006).

17. Levy, "Is the Environment a National Security Issue?"

18. Nicolas Stern. *The Economics of Climate Change. The Stern Review* (London: HM Treasury, 2006), xvi.

19. See Chapter 6 for a more detailed discussion.

20. James E. Hansen, "Letter to Prime Minister Yasuo Fukuda" (Kintnersville, PA: 2008).

21. Dabelko, "Planning for Climate Change"; Joachim von Braun and Ruth Mein-zen-Dick, *"Land-Grabbing" by Foreign Investors in Developing Countries: Risks and Opportunities* (Washington, DC: International Food Policy Research Institute, 2009).

CHAPTER 1: IMPACT OF CLIMATE CHANGE ON SECURITY

1. Jeffrey Sachs, *Common Wealth: Economics for a Crowded Planet* (New York: Penguin Press, 2008), 83.

2. James E. Hansen, "Letter to Prime Minister Yasuo Fukuda" (Kintnersville, PA: 2008).

3. Intergovernmental Panel on Climate Change, *Climate Change 2007: Synthesis Report* (IPCC, 2007).

4. Ibid., 30.

5. Ibid.

6. Ibid.

7. The IPCC uses the following categories (using expert judgment and statistical evidence) to express the assessed probability of occurrence: virtually certain >99 percent; extremely likely >95 percent; very likely >90 percent; likely >66 percent; more likely than not >50 percent; about as likely as not 33 percent to 66 percent; unlikely <33 percent; very unlikely <10 percent; extremely unlikely <5 percent; exceptionally unlikely <1 percent.

8. Intergovernmental Panel on Climate Change, *Climate Change 2007: Synthesis Report*, 30.

9. Sachs, *Common Wealth: Economics for a Crowded Planet*, 83.

10. Nicolas Stern, *The Economics of Climate Change. The Stern Review* (London: HM Treasury, 2006), 6.

11. Intergovernmental Panel on Climate Change, *Climate Change 2007: Synthesis Report*, 36.

12. Ibid.

13. Stern, *The Economics of Climate Change. The Stern Review*.

14. Ibid., iii.

15. Mike Davis, *Late Victorian Holocausts: El Niño Famines and the Making of the Third World* (London: Verso, 2002).

16. Gifford H. Miller, Marilyn L. Fogel, John W. Magee, Michael K. Gagan, Simon J. Clarke, and Beverly J. Johnson, "Ecosystem Collapse in Pleistocene Australia and a Human Role in Megafaunal Extinction," *Science* 309 (2005).

17. Jared M. Diamond, *Collapse: How Societies Choose To Fail Or Succeed* (New York: Penguin, 2006); Jürgen Scheffran, "Climate Change And Security: How Is Global Warming Affecting Existing Competition for Resources And Changing International Security Priorities? A Survey of Recent Research Shows How Complex the Picture Could Become," *Bulletin of the Atomic Scientists* 64, no. 2 (2008).

18. Scheffran, "Climate Change and Security," 20.

19. John Dearing, "Climate-Human-Environment Interactions: Resolving our Past," *Climate of the Past Discussions* 2 (2006): 565.

20. Davis, *Late Victorian Holocausts*.

21. Ibid., 12.

22. Amartya Kumar Sen, "Ingredients of Famine Analysis: Availability and Entitlements," *Quarterly Journal of Economics* 96, no. 3 (1981): 461.

23. Cecil Woodham-Smith, *The Great Hunger: Ireland 1845–1849* (London: Penguin, 1991), 75.

24. Intergovernmental Panel on Climate Change, *Climate Change 2007: Impacts, Adaptation and Vulnerability. Summary for Policymakers of the Synthesis of the IPCC Fourth Assessment Report* (IPCC, 2007), 42.

25. Ibid.

26. The origins of Malthusian theory can be found in Malthus' first essay on the principle of population. Malthus writes that "population, when unchecked, increases in a geometrical ratio; and subsistence for man in an arithmetical ratio" (Malthus and Bonar 1966: 18). This means that "unchecked" population increases at a faster pace than the production of food. This argument is based on two assumptions that man cannot exist without food and "that passion between the sexes is necessary, and will remain nearly in its present state" (Malthus and Bonar 1966: 11). That these assumptions cannot be taken as serious variables in explaining overpopulation in today's world should be a matter of common sense. Nevertheless, neo-Malthusian theory is continuously being used to impose methods controlling women's fertility and to introduce birth control policies, in particular in developing countries. Malthus' argument can be better understood in the light of the origins of his theory. Malthus' essay on population was published in 1798 in the time of the French Revolution and the beginning uprising of the Irish people calling for a republic according to the French model. The Jacobin ideas questioned unequal property ownership and supported the suppressed working class in the beginning of the industrial revolution in their fight for political representation and better working conditions. Malthus insisted on the incompatibility of welfare standards and private property rights. He argued that a welfare system would subsidize the fertility of the poor, and therefore would increase their misery with food shortages. Malthus' theory reflected the concerns of the industrialists and capitalists in England, who felt restrained by inherited institutions such as the Elizabethan Poor Laws. These laws provided the poor with relief and were established

in the sixteenth century as a form of control over the poor who had been driven out of their livelihoods by enclosures and privatization of common land (Ross 1998). Those who needed cheap labor for developing industries and agricultural production welcomed the abolition of the Poor Laws. Malthus argued that subsistence was under the threat of the rising population, and Poor Laws would only widen the gap between subsistence and population growth. There is little doubt that there was enough food to feed the poor. Even in the worst famines in British history, food was available on the market but at very high costs. Ross points out the fact that between 1750 and 1850 a quarter of England's cultivated land was being transformed into private property due to the rising commercialization of the agricultural market (Ross 1998). The loss of the commons was the tragedy and not the reverse as Hardin stated (Hardin 1968).

Garrett Hardin, "The Tragedy of the Commons," *Science*, no. 162 (1968); T. Robert Malthus, *First Essay on the Principle of Population* (London: Oxford University Press, 1798); T. Robert Malthus and James Bonar, *First Essay on Population, 1798* (London, New York: Macmillan, St. Martin's Press, 1966); Eric B. Ross, *The Malthus Factor: Population, Poverty and Politics in Capitalist Development* (London: Zed, 1998).

27. The PRIO Battle Deaths Dataset, v. 2.0, available at http://www.prio.no/ CSCW/Datasets/Armed-Conflict/Battle-Deaths. Last accessed June 23, 2009. The battle deaths data include all reported killings in battles (including civilians in crossfire) between the recorded actors. These figures do not include indirect casualties due to, e.g., hunger and epidemics in the wake of conflict; nor do they include casualties from one-sided violence (genocide, ethnic cleansing, and terrorism) and criminal behavior; Bethan Lacina and Nils Petter Gleditsch, "Monitoring Trends in Global Combat: A New Dataset of Battle Deaths," *European Journal of Population* 21, no. 2–3 (2005).

28. Halvard Buhaug, Nils Petter Gleditsch, and Ole Magnus Theisen, "Climate Change, the Environment, and Armed Conflict," presented at the Annual Meeting of the American Political Science Association, Boston, MA, August 28–31, 2008, 7.

29. Helge Brunborg and Henrik Urdal, "The Demography of Conflict and Violence: An Introduction," *Journal of Peace Research* 42, no. 4 (2005): 373.

30. Virginia Page Fortna, "Does Peacekeeping Keep Peace? International Intervention and the Duration of Peace after Civil War," *International Studies Quarterly* 48, no. 2 (2004). According to the UCDP/PRIO Armed Conflict Dataset, in 2008, only 5 conflicts were coded as wars (more than 1,000 battle-related deaths) out of a total of 36 armed conflicts. Nils Petter Gleditsch, Peter Wallensteen, Mikael Eriksson, Margareta Sollenberg, and Håvard Strand, "Armed Conflict 1946–2001: A New Dataset," *Journal of Peace Research* 39, no. 5 (2002).

31. "New wars" are characterized by new parties to the conflict, or a new type of conflict, or whether the matter of discontent (over territory or political power) has changed.

32. Buhaug, Gleditsch, and Theisen, "Climate Change, the Environment, and Armed Conflict."

33. Gleditsch, Wallensteen, Eriksson, Sollenberg, and Strand, "Armed Conflict 1946–2001: A New Dataset." Only conflicts coded as "internal armed conflict" were used.

34. Paul R. Ehrlich, *The Population Bomb* (Cutchogue, NY: Buccaneer Books, 1971); Paul R. Ehrlich and Richard L. Harriman, *How to Be a Survivor* (London: Ballantine Books Ltd: Distributed by Pan Books, 1971); Barbara Ward and René

J. Dubos, *Only One Earth: The Care and Maintenance of a Small Planet: An Unofficial Report Commissioned by the Secretary-General of the United Nations Conference on the Human Environment* (London: André Deutsch, 1972); Fred Warshofsky, *Doomsday: The Science of Catastrophe* (London: Abacus, 1979).

35. Marc A. Levy, "Is the Environment a National Security Issue?" *International Security* 20, no. 2 (1995): 44.

36. Robert D. Kaplan, "The Coming Anarchy," *Atlantic Monthly* 2 (1994): 44.

37. Ibid., 46.

38. State Failure Task Force, *State Failure Task Force Report: Phase II Findings*, no. 5 (Washington, DC: Woodrow Wilson Center, 1999), 64.

39. Thomas Homer-Dixon, "The Ingenuity Gap: Can Poor Countries Adapt to Resource Scarcity?" *Population and Development Review* 21, no. 3 (1995); Thomas Homer-Dixon, *Environment, Scarcity, and Violence* (Princeton: Princeton University Press, 1999); Thomas Homer-Dixon and Jessica Blitt, eds., *Ecoviolence: Links among Environment, Population, and Security* (Oxford: Rowman & Littlefield Publishers, 1998); Thomas Homer-Dixon and Marc A. Levy, "Correspondence: Environment and Security," *International Security* 20, no. 3 (1995).

40. Wenche Hauge and Tanja Ellingsen, "The Causal Pathway to Conflict: Beyond Environmental Scarcity," *Journal of Peace Research* 35, no. 3 (1998).

41. Homer-Dixon and Blitt, *Ecoviolence*, 224.

42. Ronnie D. Lipschutz, "Environmental Conflict: A Values-Oriented Approach," in *Conflict and the Environment*, ed. Nils Petter Gleditsch (Dordrecht: Kluwer Academic, 1997), 57.

43. Stephen C. Lonergan, *The Role of Environmental Degradation in Population Displacement* (Washington, DC: Environmental Change and Security Program, Woodrow Wilson Center, 1998), 58.

44. Henrik Urdal, "People vs. Malthus: Population Pressure, Environmental Degradation and Armed Conflict Revisited," *Journal of Peace Research* 42, no. 4 (2005): 430.

45. Indra de Soysa, "Paradise Is a Bazaar? Greed, Creed, and Governance in Civil War, 1989–99," *Journal of Peace Research* 39, no. 4 (2002); Indra de Soysa, "Ecoviolence: Shrinking Pie or Honey Pot?" *Global Environmental Politics* 2, no. 4 (2002).

46. Urdal, "People vs. Malthus," 430.

47. Macartan Humphreys and Paul Richards, *Prospects and Opportunities for Achieving the MDGs in Post-conflict Countries: A Case Study of Sierra Leone and Liberia* (New York: Center on Globalization and Sustainable Development, The Earth Institute at Columbia University), 2005.

48. Helen Ware, "Demography, Migration and Conflict in the Pacific," *Journal of Peace Research* 42, no. 4 (2005).

49. Mary Tiffen, Michael Mortimore, and Francis Gichuki, *More People, Less Erosion: Environmental Recovery in Kenya* (Chichester: Wiley, 1994).

50. Ibid., 226.

51. Ester Boserup, *Population and Technology* (Oxford: Blackwell, 1981), 124.

52. Urdal, "People vs. Malthus," 420.

53. Richard A. Matthew, "Environment, Population and Conflict," *Journal of International Affairs* 56, no. 1 (2002).

54. James K. Gasana, "Natural Resource Scarcity and Violence in Rwanda," in *Conserving the Peace: Resources, Livelihoods and Security*, ed. Mark Halle, Richard Matthew, and Jason Switzer (Winnipeg, Manitoba: International Institute for Sustainable Development, 2002), 220.

55. Jeremy Lind and Kathryn Sturman, eds., *Scarcity and Surfeit: The Ecology of Africa's Conflicts* (Pretoria: Institute for Security Studies, 2002).

56. Reinhard Mutz and Bruno Schoch, *Friedensgutachten* (Münster: LIT Verlag, 1995), 98.

57. Jon Barnett, "Destabilizing the Environment-Conflict Thesis," *Review of International Studies* 26 (2000): 271.

58. Paul Richards, *Fighting for the Rain Forest: War, Youth & Resources in Sierra Leone* (Oxford, Portsmouth, NH: International African Institute with James Currey, Heinemann, 1996), 68.

59. Ibid., 18.

60. Ibrahim Abdullah, "Bush Path to Destruction: The Origin and Character of the Revolutionary United Front/Sierra Leone," *Journal of Modern African Studies* 36, no. 2 (1998): 204.

61. David William Pearce and R. Kerry Turner, *Economics of Natural Resources and the Environment* (Hemel Hempstead: Harvester Wheatsheaf, 1990), 67; Johan Galtung, *The True Worlds: A Transnational Perspective, Preferred Worlds for the 1990's* (New York: Free Press, 1980).

62. Macartan Humphreys, "Natural Resources, Conflict, and Conflict Resolution: Uncovering the Mechanisms," *Journal of Conflict Resolution* 49, no. 4 (2005).

63. Edward Miguel, Shanker Satyanath, and Ernest Sergenti, "Economic Shocks and Civil Conflict: An Instrumental Variables Approach," *Journal of Political Economy* 112, no. 4 (2004).

64. Paul Collier, *Post-Conflict Economic Recovery* (Oxford: Department of Economics, Oxford University, 2006).

65. Paul Collier, Lani Elliott, Håvard Hegre, Anke Hoeffler, Marta Reynal-Querol, and Nicholas Sambanis, *Breaking the Conflict Trap: Civil War and Development Policy* (Washington, DC: World Bank and Oxford University Press, 2003).

66. Barbara F. Walter, "Does Conflict Beget Conflict? Explaining Recurrence in Civil War," *Journal of Peace Research* 41, no. 3 (2004).

67. Fortna, "Does Peacekeeping Keep Peace?"; Walter, "Does Conflict Beget Conflict?".

68. Urdal, "People vs. Malthus," 430.

69. Nicole Detraz and Michele M. Betsill, "Climate Change and Environmental Security: For Whom the Discourse Shifts," *International Studies Perspectives* 10, no. 3 (2009).

70. UNFCCC, *Report of the Conference of the Parties on Its Eleventh Session, Held at Montreal from 28 November to 10 December 2005. Part One: Proceedings. In UN Doc. FCCC/CP/2005/5* (Bonn: United Nations Framework Convention on Climate Change, 2006).

71. UNFCCC, *Report of the Conference of the Parties on Its Thirteenth Session, Held in Bali from 3 to 15 December 2007* (Bonn: United Nations Convention on Climate Change, 2008).

72. Christian Aid, *Human Tide: The Real Migration Crisis* (London: Christian Aid, 2007), 5. Available at http://www.christianaid.org.uk/resources/policy/climate_change.aspx. Last accessed June 26, 2009.

73. Ibid., 2.

74. The CNA Corporation, *National Security and the Threat of Climate Change* (Alexandria, VA: The CNA Corporation, 2008).

75. Thomas Homer-Dixon, "Terror in the Weather Forecast," *The New York Times*, April 24, 2007.

76. Peter Schwartz and Doug Randall, *An Abrupt Climate Change Scenario and Its Implications for United States National Security* (New York: Environmental Defense Fund, 2003), 16. Available at http://www.edf.org/documents/3566_ AbruptClimateChange.pdf. Last accessed June 29, 2009.

77. Intergovernmental Panel on Climate Change, *Climate Change 2007: Impacts, Adaptation and Vulnerability*; Stern, *The Economics of Climate Change*.

78. Alexander Carius, Dennis Tänzler, and Achim Maas, *Climate Change and Security: Challenges for German Development Cooperation* (Eschborn: Deutsche Gesellschaft für Technische Zusammenarbeit (GTZ) GmbH, 2008), 189.

79. The Economist, "A New (under) Class of Travellers," *The Economist*, June 27, 2009, 67.

80. Rachel Warren, Nigel Arnell, Robert Nicholls, Peter Levy, and Jeff Price, *Understanding the Regional Impacts of Climate Change: Research Report Prepared for the Stern Review on the Economics of Climate Change* (Norwich: Tyndall Centre, 2006), 30.

81. Ibid.

82. Barry Scott Zellen, *Arctic Doom, Arctic Boom: The Geopolitics of Climate Change in the Arctic* (Santa Barbara, CA: Praeger, 2009).

CHAPTER 2: RESOURCE SCARCITY AND SECURITY IMPLICATIONS

1. Halvard Buhaug, Nils Petter Gleditsch, and Ole Magnus Theisen, "Implications of Climate Change for Armed Conflict," in *The Social Dimensions of Climate Change: Equity and Vulnerability in a Warming World*, ed. Robin Mearns and Andrew Norton (Washington, DC: World Bank Publications, 2009), 76.

2. Intergovernmental Panel on Climate Change, *Climate Change 2007: Impacts, Adaptation and Vulnerability. Summary for Policymakers of the Synthesis of the IPCC Fourth Assessment Report* (IPCC, 2007).

3. United Nations Development Programme, *Human Development Report 2007/ 2008, Fighting Climate Change: Human Solidarity in a Divided World* (New York: UNDP, Palgrave Macmillan, 2007).

4. Alexander Carius, Dennis Tänzler, and Achim Maas, *Climate Change and Security: Challenges for German Development Cooperation* (Eschborn: Deutsche Gesellschaft für Technische Zusammenarbeit (GTZ) GmbH, 2008), 12.

5. Nicolas Stern, *The Economics of Climate Change. The Stern Review* (London: HM Treasury, 2006); German Advisory Council on Global Change, *Welt im Wandel: Sicherheitsrisiko Klimawandel* (Berlin: WBGU, 2007).

6. The CNA Corporation, *National Security and the Threat of Climate Change* (Alexandria, VA: The CNA Corporation, 2008).

7. Janet Larsen, *Setting the Record Straight: More than 52,000 Europeans Died from Heat in Summer 2003* (Earth Policy Institute, 2006). Available at http://www. earth-policy.org/Updates/2006/Update56.htm. Last accessed January 16, 2009.

8. Philippe Pirard, Stéphanie Vandentorren, Mathilde Pascal, Karine Laaidi, Alain Le Tertre, Sylvie Cassadou, and Martine Ledrans, "Summary of the Mortality Impact Assessment of the 2003 Heat Wave in France," *Eurosurveillance* 10,

no. 7 (2005). Available at http://www.eurosurveillance.org/ViewArticle.aspx?ArticleId=554. Last accessed January 16, 2009.

9. Ibid.

10. David S. Battisti and Rosamond L. Naylor, "Historical Warnings of Future Food Insecurity with Unprecedeted Seasonal Heat," *Science* 323 (2009).

11. Ibid.

12. Ibid.

13. Ibid.

14. Serigne Kandji, Louis Verchot, and Jens Mackensen, *Climate Change and Variability in the Sahel Region: Impacts and Adaptation Strategies in the Agricultural Sector*. Nairobi: United Nations Environmental Programme and World Agroforestry Center, 2006.

15. Marshall B. Burke, Edward Miguel, Shanker Satyanath, John A. Dykema, and David B. Lobell, "Warming Increases the Risk of Civil War in Africa," *Proceedings of the National Academy of Sciences* 106, no. 49 (2009). Note: This study only considers wars (more than 1,000 battle-related deaths) without including small-scale conflicts (more than 25 battle-related deaths) that are believed to be even more relevant to climate change-induced conflict.

16. Intergovernmental Panel on Climate Change, *Climate Change 2007: Impacts, Adaptation and Vulnerability*.

17. United Nations Population Division, *World Population Prospects: The 2002 Revision*. New York: United Nations Population Division, 2003.

18. Intergovernmental Panel on Climate Change, *Climate Change 2007: Impacts, Adaptation and Vulnerability*.

19. Casey Brown and Upmanu Lall, "Water and Economic Development: The Role of Variability and a Framework for Resilience," *Natural Resources Forum* 30, no. 4 (2006).

20. Bates Bryson, Zbigniew W. Kundzewicz, Shaohong Wu, and Jean Palutikof, *Climate Change and Water* (Nairobi: Intergovernmental Panel on Climate Change (IPCC), 2008), 81.

21. Carius, Tänzler, and Maas, *Climate Change and Security*, 20.

22. Nigel W. Arnell, "Climate Change and Global Water Resources: SRES Emissions and Socio-Economic Scenarios," *Global Environmental Change* 14, no. 1 (2004).

23. Peter J. Ashton, "Avoiding Conflicts over Africa's Water Resources" *Ambio* 31, no. 3 (2002): 238.

24. James K. A. Benhin, *Climate Change and South African Agriculture: Impacts and Adaptation Options* (Pretoria: The Centre for Environmental Economics and Policy in Africa, University of Pretoria, 2006).

25. Jeffrey D. Sachs, "The Strategic Significance of Global Inequality," *The Washington Quarterly* 24, no. 3 (2001), 187.

26. World Bank, *World Development Indicators* (Washington, DC: The World Bank, 2007).

27. United Nations Economic Commission for Africa, *Economic Report on Africa* (Addis Ababa: UNECA, 2005).

28. Carius, Tänzler, and Maas, *Climate Change and Security*, 8.

29. James D. Fearon and David D. Laitin, "Ethnicity, Insurgency and Civil War," *American Political Science Review* 97, no. 1 (2003).

30. Ragnhild Nordås and Nils Petter Gleditsch, "Climate Change and Conflict," *Political Geography* 26, no. 6 (2007).

31. Cullen S. Hendrix and Sarah M. Glaser, "Trends and Triggers: Climate, Climate Change and Civil Conflict in sub-Saharan Africa," *Political Geography* 26, no. 6 (2007).

32. Patrick Meier, Doug Bond, and Joe Bond, "Environmental Influences on Pastoral Conflict in the Horn of Africa," *Political Geography* 26, no. 6 (2007).

33. Stephan Faris, "The Real Roots of Darfur," *Atlantic Monthly*, 10 April 2007. Available at http://www.theatlantic.com/doc/200704/darfur-climate. Last accessed September 2, 2009.

34. United Nations Environment Programme, *Sudan Post-Conflict Environmental Assessment* (Nairobi: UNEP, 2007), 8.

35. Mike Davis, *Late Victorian Holocausts: El Niño Famines and the Making of the Third World* (London: Verso, 2002).

36. David D. Zhang, Jane Zhang, Harry F. Lee, and Yuan-qing He, "Climate Change and War Frequency in Eastern China over the Last Millennium," *Human Ecology* 35 (2007).

37. Clionadh Raleigh and Henrik Urdal, "Climate Change, Environmental Degradation and Armed Conflict," *Political Geography* 26, no. 6 (2007); Ole Magnus Theisen, Helge Holtermann, and Halvard Buhaug, "Drought, Political Exclusion, and Civil War," presented at the *International Studies Association annual convention* (New Orleans, LA, 17–20 February: 2010).

38. Jon Barnett and W. Neil Adger, "Climate Change, Human Security and Violent Conflict," *Political Geography* 26, no. 6 (2007).

39. Paul Collier and Anke Hoeffler, "On the Incidence of Civil War in Africa," *Journal of Conflict Resolution* 46, no. 1 (2002); Ibrahim Elbadawi and Nicholas Sambanis, "How Much War Will We See? Explaining the Prevalence of Civil War," *Journal of Conflict Resolution* 46, no. 3 (2002); Fearon and Laitin, "Ethnicity, Insurgency and Civil War."

40. Edward Miguel, Shanker Satyanath, and Ernest Sergenti, "Economic Shocks and Civil Conflict: An Instrumental Variables Approach," *Journal of Political Economy* 112, no. 4 (2004).

41. I use a data set developed by the Department of Peace and Conflict Research, Uppsala University and Centre for the Study of Civil War at the International Peace Research Institute, Oslo (referred to as UCDP/PRIO armed conflict dataset). The UCDP/PRIO armed conflict dataset reports all conflicts that are above a certain threshold, namely more than 25 reported battle-related deaths per year. The data are divided into four types of conflicts: extrasystemic armed conflict, interstate armed conflict, internal armed conflict, and internationalized internal armed conflict.

Nils Petter Gleditsch, Peter Wallensteen, Mikael Eriksson, Margareta Sollenberg, and Håvard Strand, "Armed Conflict 1946–2001: A New Dataset," *Journal of Peace Research* 39, no. 5 (2002).

42. Drawn from Fearon and Laitin, "Ethnicity, Insurgency and Civil War."

43. Macartan Humphreys, "Natural Resources, Conflict, and Conflict Resolution: Uncovering the Mechanisms," *Journal of Conflict Resolution* 49, no. 4 (2005): 513.

44. Fearon and Laitin, "Ethnicity, Insurgency and Civil War."

45. John Drysdale, *Stoics Without Pillows: A Way Forward for the Somalilands* (London: HAAN Associates Publishing, 2000), 2.

46. Jeanne X. Kasperson, Roger E. Kasperson, and Billie L. Turner II, eds., *Regions at Risk: Comparisons of Threatened Environments* (Tokyo: United Nations University Press, 1995).

47. Peter Conze and Thomas Labahn, "From a Socialistic System to a Mixed Economy: The Changing Framework for Somali Agriculture," in *Somalia: Agriculture in the Winds of Change*, ed. Peter Conze and Thomas Labahn (Saarbrücken: EPI Verlag, 1986).

48. Ben Wisner, "Jilaal, Gu, Hagaa, and Der: Living with the Somali Land, and Living Well," in *The Somali Challenge: From Catastrophe to Renewal?* Ed. Ahmed I. Samatar (Boulder, London: Lynne Rienner, 1994).

49. Drysdale, *Stoics without Pillows: A Way Forward for the Somalilands*.

50. Peter D. Little, *Somalia: Economy without State* (Oxford, Bloomington & Indianapolis, Hargeisa: James Currey, Indiana University Press, Btec Books, 2003), 65.

51. Bernhard Helander, "The Hubeer in the Land of Plenty: Land, Labor, and Vulnerability Among a Southern Somali Clan," in *The Struggle for Land in Southern Somalia: The War Behind the War*, ed. Catherine Lowe Besteman and Lee V. Cassanelli (London, Boulder: HAAN Associates Publishing, Westview Press, 1996), 48.

52. United Nations Development Programme, *Human Development Report, Somalia 2001* (Nairobi: United Nations Development Programme Somalia Country Office, 2001), 74.

53. Christian Webersik, "Wars over Resources? Evidence from Somalia," *Environment* 50, no. 3 (2008).

54. Christian Webersik, "Differences That Matter: The Struggle of the Marginalised in Somalia," *Africa* 74, no. 4 (2004).

55. Mats R. Berdal, David Malone, and International Peace Academy, *Greed & Grievance: Economic Agendas in Civil Wars* (Boulder, London, Ottawa: Lynne Rienner Publishers, International Development Research Centre, 2000); David Keen, "Incentives and Disincentives for Violence," in *Greed & Grievance: Economic Agendas in Civil Wars*, ed. Mats R. Berdal, David Malone, and International Peace Academy (Boulder, London, Ottawa: Lynne Rienner Publishers, International Development Research Centre, 2000).

56. Alexander de Waal, *Famine Crimes: Politics and the Disaster Relief Industry in Africa, African Issues* (Oxford, Bloomington: Currey, Indiana University Press, 1997), 159.

57. United Nations Environment Programme, *Sudan Post-Conflict Environmental Assessment*, 8.

58. Ibid.

59. Alemayehu Kassa, "Drought Risk Monitoring for the Sudan" (SOAS, 1999).

60. United Nations Environment Programme, *Sudan Post-Conflict Environmental Assessment*, 93.

61. Stephen Stedman, "Spoiler Problems in Peace Processes," *International Security* 22, no. 2 (1997).

62. Macartan Humphreys, Jeffrey Sachs, and Joseph E. Stiglitz, *Escaping the Resource Curse* (New York: Columbia University Press, 2007).

63. Climate Change Network Nepal, "Climate Change and its Impact in Nepal," *CCNN Newsletter* Year 1, no. 1 (2007): 2.

64. Marc A. Levy, Catherine Thorkelson, Charles Vörösmarty, Ellen Douglas, and Macartan Humphreys, "Freshwater Availability Anomalies and Outbreak of Internal War: Results from a Global Spatial Time Series Analysis," presented at the *Human Security and Climate Change International Workshop*, Holmen Fjord Hotel, Asker, near Oslo, 21–23 June (2005).

65. S. Mansoob Murshed and Scott Gates, "Spatial-Horizontal Inequality and the Maoist Insurgency in Nepal," *Review of Development Economics* 9, no. 1 (2005).

66. Manish Thapa, "Maoist Insurgency in Nepal: Context, Cost and Consequences," in *Afro-Asian Conflicts*, ed. Seema Shekhawat and Debidatta Aurobinda Mahapatra (New Delhi: New Century Publications, 2008), 80.

67. International Centre for Integrated Mountain Development, "Flash Floods in the Himalayas," *ICIMOD* (2008).

68. World Wide Fund for Nature, *Glaciers, Glacier Retreat, and its Subsequent Impacts in Nepal, India and China*. Kathmandu: WWF Nepal Program, 2005, 25; Shardul Agrawala, *Bridge over Troubled Waters: Linking Climate Change and Development* (Paris: OECD, 2005).

69. Arun B. Shrestha, Cameron P. Wake, Paul A. Mayewski, and Jack E. Dibb., "Maximum Temperature Trends in the Himalaya and Its Vicinity: An Analysis Based on Temperature Records from Nepal for the Period 1971–94," *Journal of Climate* 12 (1999).

70. Interview with Andreas Schild, Director General of ICIMOD in Mountain Forum, "Natural Resources: Women, Conflicts and Management," *Mountain Forum Bulletin* 8, no. 2 (2008): 30.

71. Ibid.

72. Theisen, Holtermann, and Buhaug, "Drought, Political Exclusion, and Civil War."

73. Intergovernmental Panel on Climate Change, *Climate Change 2007: Impacts, Adaptation and Vulnerability*.

CHAPTER 3: NATURAL DISASTERS AND SECURITY IMPLICATIONS

1. Alexander Carius, Dennis Tänzler, and Achim Maas, *Climate Change and Security: Challenges for German Development Cooperation* (Eschborn: Deutsche Gesellschaft für Technische Zusammenarbeit (GTZ) GmbH, 2008), 15.

2. A disaster is listed in CRED's Emergency Events Database (EM-DAT) when at least one of the following criteria is met: 10 or more people killed, 100 or more people affected, declaration of a state of emergency, or a call for international assistance. See more at www.emdat.be. Last accessed February 2, 2009.

3. Halvard Buhaug, Nils Petter Gleditsch, and Ole Magnus Theisen, "Climate Change, the Environment, and Armed Conflict," presented at the *Annual Meeting of the American Political Science Association* (Boston, MA, August 28–31, 2008), 6.

4. Ibid.

5. Ibid.

6. Maxx Dilley, Robert S. Chen, Uwe Deichmann, Arthur L. Lerner-Lam, and Margaret Arnold, *Natural Disaster Hotspots: A Global Risk Analysis, Synthesis Report* (Washington, DC: International Bank for Reconstruction and Development/The World Bank and Columbia University), 2005.

7. Buhaug, Gleditsch, and Theisen, "Climate Change, the Environment, and Armed Conflict," 6.

8. The CNA Corporation, *National Security and the Threat of Climate Change* (Alexandria, VA: The CNA Corporation, 2008), 3.

9. Ibid., 20.

10. William D. Nordhaus, *The Economics of Hurricanes in the United States* (Boston: Annual Meetings of the American Economic Association, 2006), 4.

11. Peter Webster, Greg Holland, Judith A. Curry, and H.-R. Chang, "Changes in Tropical Cyclone Number, Duration, and Intensity in a Warming Environment," *Science* 309, no. 5742 (2005): 1846.

12. Christopher W. Landsea, Bruce A. Harper, Karl Hoarau, and John A. Knaff, "Can We Detect Trends in Extreme Tropical Cyclones?" *Science* 313, no. 5786 (2006): 452.

13. Webster, Holland, Curry, and Chang, "Changes in Tropical Cyclone Number, Duration, and Intensity in a Warming Environment," 1844.

14. Paula A. Agudelo and Judith A. Curry, "Analysis of Spatial Distribution in Tropospheric Temperature Trends," *Geophysical Research Letters* 31, no. L22207, DOI:10.1029/2004GL020818 (2004): 4.

15. Charles Ehrhart, Andrew Thow, Mark de Blois, and Alyson Warhurst, *Humanitarian Implications of Climate Change: Mapping Emerging Trends and Risk Hotspots* (UN Office for the Coordination of Humanitarian Affairs and CARE International, 2008), 11.

16. Dilley, Chen, Deichmann, Lerner-Lam, and Arnold, *Natural Disaster Hotspots*, 23.

17. Resilience in ecology measures the capability of an ecosystem to cope with external stress and environmental change, and to the ability to recover after an external shock.

18. W. Neil Adger and Timothy O'Riordan, "Population, Adaptation and Resilience," in *Environmental Science for Environmental Management*, ed. Timothy O'Riordan (Harlow: Prentice Hall, Pearson Education Limited, 1999).

19. Ibid.

20. Dawn Chatty, *Mobile Pastoralists: Development Planning and Social Change in Oman* (New York: Columbia University Press, 1996).

21. Adger and O'Riordan, "Population, Adaptation and Resilience."

22. Mick Kelly and Neil Adger, *Assessing Vulnerability to Climate Change and Facilitating Adaptation.* (Norwich: Centre for Social and Economic Research on the Global Environment, University of East Anglia, Norwich, and University College London, 1999).

23. Piers Blaikie, Terry Cannon, Ian Davis, and Ben Wisner, eds., *At Risk: Natural Hazards, People's Vulnerability, and Disasters* (London: Routledge, 1994).

24. Poor people as well as rich people might live in areas that are prone to natural disasters, but the poor do not have the choice to move into safer areas. Rich people may live at the top of steep valleys for the "nice view."

25. James S. Coleman, "Social Capital in the Creation of Human Capital," *American Journal of Sociology, Supplement Organizations and Institutions: Sociological and Economic Approaches to the Analysis of Social Structure* 94 (1988).

26. Ben Fine, "Social Capital?" *Development and Change* 30, no. 1 (1999).

27. William M. Gray, "Global View of the Origin of Tropical Disturbances and Storms," *Monthly Weather Review* 96, no. 10 (1968).

28. Ann Henderson-Sellers, Hao Zhang, Gerhard Berz, Kerry Emanuel, William Gray, Christopher W. Landsea, Greg Holland, et al., "Tropical Cyclones and Global Climate Change: A Post-IPCC Assessment," *Bulletin of the American Meteorological Society* 79, no. 1 (1998): 19.

29. Lloyd J. Shapiro and Stanley B. Goldenberg, "Atlantic Sea Surface Temperatures and Tropical Cyclone Formation," *Journal of Climate* 11, no. 4 (1998).

30. Webster, Holland, Curry, and Chang, "Changes in Tropical Cyclone Number, Duration, and Intensity in a Warming Environment," 1846.

31. Nicolas Stern, *The Economics of Climate Change. The Stern Review* (London: HM Treasury, 2006), 89.

32. M. M. Kabir, Bibhas Chandra Saha, and Jalauddin M. Abdul Hye, *Cyclonic Storm Surge Modelling for Design of Coastal Polder* (Dhaka, Bangladesh: Institute of Water Modelling, 2009). Available at www.iwmbd.org/html/PUBS/publications/P024.PDF. Last accessed January 24, 2009.

33. This figure reflects the insured market loss. It does not include the flood and storm-surge losses that are covered under the National Flood Insurance Program (NFIP). Munich Re Group, *Katrina and Rita: Munich Re Estimates Total Insured Market Losses at up to US$40bn.* Munich: Munich Re Group, 2005. Available at http://www.munichre.com/en/press/press_releases/2005/2005_09_28_press_release.aspx. Last accessed July 25, 2008.

34. Munich Re Group, *Topics Geo: Natural Catastrophes 2006* (Munich: Munich Re Group, 2007), 45.

35. Filippo Giorgi, Bruce Hewitson, J. Christensen, Michael Hulme, Hans Von Storch, Penny Whetton, R. Jones, et al., "Regional Climate Information-Evaluation and Projections," in *Climate Change 2001: The Scientific Basis*, ed. John T. Houghton, Y. Ding, D. J. Griggs, and M. Noguer. (Cambridge, UK: Cambridge University Press, 2001).

36. Laurens M. Bouwer, Ryan P. Crompton, Eberhard Faust, Peter Höppe, and Roger A. Pielke Jr., "Confronting Disaster Losses," *Science* 318, no. 2 (2007): 753.

37. Munich Re Group, *Topics Geo: Natural Catastrophes 2006*, 45.

38. Ibid.

39. Ibid.

40. Nordhaus, *The Economics of Hurricanes in the United States*.

41. Christian Webersik, Miguel Esteban, and Tomoya Shibayama, "The Economic Impact of Future Increase in Tropical Cyclones in Japan," *Natural Hazards*, DOI: 10.1007/s11069-010-9522-9 (2010); Miguel Esteban, Christian Webersik, and Tomoya Shibayama, "Effect of a Global Warming-Induced Increase in Typhoon Intensity on Urban Productivity in Taiwan," *Sustainability Science* 4, no. 2 (2009).

42. Much of the following methodology to predict future increase of tropical cyclones is based on Miguel Esteban, Christian Webersik, and Tomoya Shibayama, "Methodology for the Estimation of the Increase in Time Loss due to Future Increase in Tropical Cyclone Intensity in Japan," *Climatic Change*, DOI 10.1007/s10584-009-9725-9 (2009).

43. Stern, *The Economics of Climate Change*.

44. Clionadh Raleigh and Henrik Urdal, "Climate Change, Environmental Degradation and Armed Conflict," *Political Geography* 26, no. 6 (2007).

45. Esteban, Webersik, and Shibayama, "Methodology for the Estimation of the Increase in Time Loss due to Future Increase in Tropical Cyclone Intensity in Japan."

46. Webersik, Esteban, and Shibayama, "The Economic Impact of Future Increase in Tropical Cyclones in Japan."

47. Bouwer, Crompton, Faust, Höppe, and Pielke Jr., "Confronting Disaster Losses."

48. Webersik, Esteban, and Shibayama, "The Economic Impact of Future Increase in Tropical Cyclones in Japan."

49. Intergovernmental Panel on Climate Change, *Climate Change 2007: Impacts, Adaptation and Vulnerability. Summary for Policymakers of the Synthesis of the IPCC Fourth Assessment Report* (IPCC, 2007).

50. Thomas R. Knutson, Robert E. Tuleya, Weixing Shen, and Isaac Ginis, "Impact of CO_2-Induced Warming on Hurricane Intensities as Simulated in a Hurricane Model with Ocean Coupling," *Journal of Climate* 14 (2001).

51. Stern, *The Economics of Climate Change*, vi.

52. Ibid., 92.

53. Kerry A. Emanuel, "The Dependence of Hurricane Intensity on Climate," *Nature* 326, no. 6112 (1987).

54. Nordhaus, *The Economics of Hurricanes in the United States.*

55. Stern, *The Economics of Climate Change*, 92.

56. Emanuel, "The Dependence of Hurricane Intensity on Climate."

57. Bouwer, Crompton, Faust, Höppe, and Pielke Jr., "Confronting Disaster Losses," 753.

58. Jared M. Diamond, *Collapse: How Societies Choose to Fail or Succeed* (New York: Penguin, 2006).

59. Ibid., 333.

60. Intergovernmental Panel on Climate Change, *Climate Change 2007*, 30.

61. For drought, the authors used the Weighted Anomaly of Standardized Precipitation (WASP) developed by the International Research Institute for Climate and Society (IRI) at Columbia University, computed on a 2.5° x 2.5° grid from monthly average precipitation data for 1980–2000. The WASP assesses the degree of precipitation deficit or surplus over a specified number of months, weighted by the magnitude of the seasonal cyclic variation in precipitation. A three-month running average was applied to the precipitation data and the median rainfall for the 21-year period calculated for each grid point. A mask was applied to eliminate grid points where the three-month running average precipitation was less than one mm per day. This excluded both desert regions and dry seasons from the analysis. For the remaining points, a drought event was identified when the magnitude of a monthly precipitation deficit was less than or equal to 50 percent of its long-term median value for three or more consecutive months. Dilley, Chen, Deichmann, Lerner-Lam, and Arnold, *Natural Disaster Hotspots*, 5.

62. Stern, *The Economics of Climate Change*, 74.

63. Ibid., 80.

64. Ibid., 84.

65. Alexander de Waal, *Famine Crimes: Politics and the Disaster Relief Industry in Africa, African Issues* (Oxford, Bloomington: Currey, Indiana University Press, 1997), 159.

66. Amartya Kumar Sen, "Food, Economics and Entitlements," in *The Political Economy of Hunger: Selected Essays*, ed. Jean Drèze, Amartya Kumar Sen, and Athar Hussain (Oxford: Clarendon Press, 1995).

67. Amartya Kumar Sen, "Ingredients of Famine Analysis: Availability and Entitlements," *Quarterly Journal of Economics* 96, no. 3 (1981): 434.

68. T. Robert Malthus, *First Essay on the Principle of Population* (London: Oxford University Press, 1798).

69. Sen, "Food, Economics and Entitlements," 51.

70. Ibid.

71. Stephen Devereux, *Theories of Famine* (New York, London: Harvester Wheatsheaf, 1993).

72. Mohamed Haji Mukhtar, "The Plight of the Agro-Pastoral Society of Somalia," *Review of African Political Economy* 23, no. 70 (1996).

73. Steven Hansch, Scott Lillibridge, Grace Egeland, Charles Teller, and Michael Toole, *Lives Lost, Lives Saved: Excess Mortality and the Impact of Health Interventions in the Somalia Emergency* (Washington, DC: Refugee Policy Group, 1994), 24.

74. Helmut Opletal, "'Nicht der Krieg schuf das Desaster in Somalia' Ein Bericht von Africa Watch nennt die Ursachen für Hunger und Krieg am Horn von Afrika," *Frankfurter Rundschau*, May 14, 1993.

75. Africa Watch and Physicians for Human Rights, "Somalia: No Mercy in Mogadishu: The Human Cost of the Conflict & the Struggle for Relief," *Africa Watch and Physicians for Human Rights* 4, no. 4 (1992): 19.

76. Ibid.

77. Refugee Policy Group, *Hope Restored? Humanitarian Aid in Somalia, 1990–1994* (Washington, DC: Refugee Policy Group), 1994.

78. Buhaug, Gleditsch, and Theisen, "Climate Change, the Environment, and Armed Conflict," 12.

79. Aaron T. Wolf, "Conflict and Cooperation along International Waterways," *Water Policy* 1, no. 1 (1998); Aaron T. Wolf, "Water Wars and Water Reality: Conflict and Cooperation along International Waterways," in *Environmental Change, Adaptation and Human Security*, ed. Steve Lonergan (Dordrecht: Kluwer, 1999).

80. Nils Petter Gleditsch, Kathryn Furlong, Håvard Hegre, Bethany Lacina, and Taylor Owen, "Conflicts over Shared Rivers: Resource Scarcity or Fuzzy Boundaries?" *Political Geography* 25, no. 4 (2006).

81. Marit Brochmann and Nils Petter Gleditsch, "Shared Rivers and International Cooperation," in paper presented at the Workshop on Polarization and Conflict, Nicosia, Cyprus, 26–29 April (2006).

82. Buhaug, Gleditsch, and Theisen, "Climate Change, the Environment, and Armed Conflict," 13.

83. Jeffrey D. Sachs, *Common Wealth: Economics for a Crowded Planet* (New York: Penguin Press, 2008).

84. Intergovernmental Panel on Climate Change, *Climate Change 2007: Impacts, Adaptation and Vulnerability*; Stern, *The Economics of Climate Change*.

85. Esteban, Webersik, and Shibayama, "Methodology for the Estimation of the Increase in Time Loss due to Future Increase in Tropical Cyclone Intensity in Japan."

86. Miguel Esteban and Christian Webersik, "The Cost of Inaction: Estimation of the Economic Effects of an Increase in Typhoon Intensity in the Asia-Pacific Region," *International Journal of Global Warming* (2010 forthcoming).

87. Esteban, Webersik, and Shibayama, "Methodology for the Estimation of the Increase in Time Loss due to Future Increase in Tropical Cyclone Intensity in Japan."

88. Per Stromberg, Miguel Esteban, and Dexter Thompson-Pomeroy. *Interlinkages in Climate Change: Vulnerability of a Mitigation Strategy? Impact of Increased Typhoon Intensity on Biofuel Production in the Philippines*. Tokyo, 2009.

89. Theisen, Holtermann, and Buhaug, "Drought, Political Exclusion, and Civil War."

CHAPTER 4: MIGRATION AS AN OUTCOME

1. According to the U.S. Census Bureau, available at http://www.census.gov/main/www/popclock.html. Last accessed September 20, 2009.

2. Clionadh Raleigh and Lisa Jordan, "Climate Change, Migration, and Conflict," presented at the *International Studies Association annual convention* (San Francisco, 26–29 March: 2007), 3.

3. Timothy M. Lenton, Hermann Held, Elmar Kriegler, Jim W. Hall, Wolfgang Lucht, Stefan Rahmstorf, and Hans Joachim Schellnhuber, "Tipping Elements In The Earth's Climate System," *Proceedings of the National Academy of Sciences* 105, no. 6 (2008).

4. The IPCC uses precisely defined language to describe likelihoods. Virtually certain > 99% probability of occurrence, very likely = 90–99%, likely = 66–89%, about as likely as not = 33–66%, unlikely = 10–33%, very unlikely = 1–10%, and exceptionally unlikely = < 1% probability.

5. Intergovernmental Panel on Climate Change, *Climate Change 2007: Impacts, Adaptation and Vulnerability. Summary for Policymakers of the Synthesis of the IPCC Fourth Assessment Report* (IPCC, 2007), 18.

6. EM-DAT, "Emergency Events Database," http://www.emdat.be.

7. Alexander de Waal, *Famine Crimes: Politics and the Disaster Relief Industry in Africa, African Issues* (Oxford, Bloomington: Currey, Indiana University Press, 1997).

8. EM-DAT, "Emergency Events Database."

9. Intergovernmental Panel on Climate Change, *Climate Change 2001: Impacts, Adaptation, and Vulnerability. Contribution of Working Group II to the Third Assessment Report of the Intergovernmental Panel on Climate Change* (Cambridge: Cambridge University Press, 2001); Sabine Perch-Nielsen, "Understanding the Effect of Climate Change on Human Migration: The Contribution of Mathematical and Conceptual Models" (Zurich: Swiss Federal Institute of Technology, 2004).

10. Intergovernmental Panel on Climate Change, *Climate Change 2001: The Scientific Basis. Contribution of Working Group I to the Third Assessment Report of the Intergovernmental Panel on Climate Change* (Cambridge: Cambridge University Press, 2001).

11. Raleigh and Jordan, "Climate Change, Migration and Conflict," 7.

12. Diane C. Bates, "Environmental Refugees? Classifying Human Migrations Caused by Environmental Change," *Population and Environment* 23, no. 5 (2002): 465.

13. Frank Biermann and Ingrid Boas, "Protecting Climate Refugees: The Case for a Global Protocol," *Environment* 50, no. 6 (2008); Richard Black, *Environmental Refugees: Myth or Reality?* (Geneva: UNHCR, 2001); Bo R. Doos, "Can Large-Scale Environmental Migrations Be Predicted?" *Global Environmental Change* 7 (1997); Lori M. Hunter, "Migration and Environmental Hazards," *Population and Environment* 26, no. 4 (2005); Stephen C. Lonergan, *The Role of Environmental Degradation in Population Displacement* (Washington, DC: Environmental Change and Security Program, Woodrow Wilson Center, 1998); Henrik Urdal, "People vs. Malthus: Population Pressure, Environmental Degradation and Armed Conflict Revisited," *Journal of Peace Research* 42, no. 4 (2005).

14. Norman Myers, *Environmental Exodus: An Emergent Crisis in the Global Arena* (Washington, DC: Climate Institute, 1995), Norman Myers, "Environmental Refugees," *Population and Environment: A Journal of Interdisciplinary Studies* 19 (1997). For a critique of Norman Myers' thesis calling for a more cautious approach, see Raleigh and Jordan, "Climate Change, Migration and Conflict."

15. Essam El-Hinnawi, *Environmental Refugees* (Nairobi: United Nations Environmental Programme, 1985), 4.

16. Susana B. Adamo and Alexander de Sherbinin, *The Impact of Climate Change on the Spatial Distribution of Populations and Migration*, *A Report prepared for the United Nations Population Division* (Palisades: Center for International Earth Science Information Network (CIESIN), The Earth Institute at Columbia University, 2008), 32.

17. Graeme Hugo, "Environmental Concerns and International Migration," *International Migration Review* 30 (1996); Bates, "Environmental Refugees? Classifying Human Migrations Caused by Environmental Change."

18. Bates, "Environmental Refugees?" 469.

19. Raleigh and Jordan, "Climate Change, Migration and Conflict," 3.

20. Christian Klose and Christian Webersik, "Assessing the Inter-Relationship between Geohazards, Sustainable Development, and Political Stability: A Comparative Study for Haiti and the Dominican Republic," presented at the *7th International Science Conference on the Human Dimensions of Global Environmental Change (Open Meeting 2009)* (Bonn: 2009).

21. World Bank, *Natural Disaster Hotspots: A Global Risk Analysis* (Washington, DC: World Bank, 2005).

22. Intergovernmental Panel on Climate Change, *Climate Change 2007: Impacts, Adaptation and Vulnerability*; Stern, *The Economics of Climate Change*, 35.

23. United Nations Environment Programme, *Marine and Coastal Ecosystems and Human Well-Being: A Synthesis Report Based on the Findings of the Millennium Ecosystem Assessment* (Nairobi: United Nations Environment Programme, 2006).

24. Raleigh and Jordan, "Climate Change, Migration and Conflict," 19.

25. Adamo and de Sherbinin, *The Impact of Climate Change on the Spatial Distribution of Populations and Migration*.

26. Gordon McGranahan, Deborah Balk, and Bridget Anderson, "The Rising Tide: Assessing the Risks of Climate Change and Human Settlements in Low Elevation Coastal Zones," *Environment and Urbanization* 19, no. 1 (2007): 19.

27. The following analysis is based on Raleigh and Jordan, "Climate Change, Migration and Conflict." Also see EM-DAT, "Emergency Events Database."

28. Alexander de Sherbinin, "A 'Human' Perspective on Four Deltas," in *GWSP-LOICZ-CSDMS Scoping Workshop on Dynamics and Vulnerability of River Delta Systems*, Boulder, Colorado, 26–28 September (2007).

29. Jeffrey D. Sachs, *Common Wealth: Economics for a Crowded Planet* (New York: Penguin Press, 2008), 163.

30. Adamo and de Sherbinin, *The Impact of Climate Change on the Spatial Distribution of Populations and Migration*, 24.

31. The use of the term *climate change victim* has been informed by ongoing discussions related to a joint-research project on "Victimization in a Climate

Constrained World" of the United Nations University/Institute for Sustainability and Peace and Tokiwa University/Tokiwa International Institute of Victimology (TIVI).

32. McGranahan, Balk, and Anderson, "The Rising Tide."

33. Raleigh and Jordan, "Climate Change, Migration and Conflict," 21.

34. Population Division of the Department of Economic and Social Affairs of the United Nations Secretariat, *World Population Prospects: The 2005 Revision* (New York, 2005); Population Division of the Department of Economic and Social Affairs of the United Nations Secretariat, *World Population Prospects: The 2006 Revision* (New York, 2006).

35. More recent figures are unavailable. The study integrates recently developed spatial databases of finely resolved global population distribution, urban extents, and elevation data to produce country-level estimates of urban land area and population in low-elevation coastal zones (LECZs). The data set is complex, but the basic approach is simple: by overlaying geographic data layers, we calculate the population and land area in each country, in its LECZ, and then summarize by country, region, and economic grouping. Three geographic data sets are required. The first delineates the extent of the LECZ itself. We defined the LECZ as land area contiguous with the coastline up to a 10-meter rise elevation. The second required data set is one that delineates urban footprints. Urban extents are based largely on the NOAA's nighttime light "city lights" 1994–95 satellite data set coupled with settlement information (e.g., name and population) to verify that the lighted area corresponds to a human settlement (rather than an industrial location, for example). The third data set required for this analysis is a population grid for each country, and associated with each grid cell, the land area. All data are expressed at one-kilometer resolution. McGranahan, Balk, and Anderson, "The Rising Tide," 21.

36. Ibid., 22.

37. Ibid., 34.

38. The Maldives, Marshall Islands, Tuvalu, Cayman Islands, and Turk and Caicos Islands, all countries with a total population of less than 100,000 people or smaller than 1,000 square kilometers, were excluded from this list. Ibid., 26.

39. Ibid., 33.

40. Ibid., 30.

41. Tata Energy Research Institute, *The Economic Impact of a One-Metre Sea-Level Rise on the Indian Coastline: Method and Case Studies, Report Submitted to the Ford Foundation* (Delhi: TERI, 1996).

42. Alexander de Sherbinin, Andrew Schiller, and Alex Pulsipher, "The Vulnerability of Global Cities to Climate Hazards," *Environment and Urbanization* 39, no. 1 (2007).

43. Saleemul Huq, Sari Kovats, Hannah Reid, and David Satterthwaite, "Editorial: Reducing Risks to Cities from Disasters and Climate Change," *Environment and Urbanization* 19, no. 1 (2007).

44. United Nations Environment Programme, *Fourth Global Environment Outlook: Environment for Development Assessment Report* (Nairobi: United Nations Environment Programme, 2007).

45. United Nations Environment Programme, *Marine and Coastal Ecosystems and Human Well-Being.*

46. Robert McLeman and Barry Smit, "Migration as an Adaptation to Climate Change," *Climatic Change* 76 (2006): 39.

47. United Nations Development Programme, *Human Development Report, Somalia 2001* (Nairobi: United Nations Development Programme Somalia Country Office, 2001).

48. Adamo and de Sherbinin, *The Impact of Climate Change on the Spatial Distribution of Populations and Migration*, 30.

49. Saul S. Morris, Oscar Neidecker-Gonzales, Calogero Carletto, Marcial Munguía, Juan Manuel Medina, and Quentin Wodon, "Hurricane Mitch and the Livelihoods of the Rural Poor in Honduras," *World Development* 30, no. 1 (2002).

50. McLeman and Smit, "Migration as an Adaptation to Climate Change," 33.

51. Sabine Henry, Victor Piché, Dieudonné Ouédraogo, and Eric F. Lambin, "Descriptive Analysis of the Individual Migratory Pathways According to Environmental Typologies," *Population and Environment* 25, no. 5 (2004).

52. M. Q. Zaman, "The Displaced Poor and Resettlement Policies in Bangladesh," *Disasters* 15, no. 2 (1991).

53. Ngai Weng Chan, "Choice and Constraints in Floodplain Occupation: The Influence of Structural Factors on Residential Location in Peninsular Malaysia," *Disasters* 19, no. 4 (1995).

54. Raleigh and Jordan, "Climate Change, Migration and Conflict."

55. Sukhan Jackson and Adrian C. Sleigh, "Resettlement for China's Three Gorges Dam: Socio-Economic Impact and Institutional Tensions," *Communist and Post-Communist Studies* 33 (2000).

56. Hunter, "Migration and Environmental Hazards," 297.

57. Ibid., 284–285.

58. Nils Petter Gleditsch, Ragnhild Nordås, and Idean Salehyan, *Climate Change and Conflict: The Migration Link: Coping with Crisis* (New York: International Peace Academy, 2007).

59. Idean Salehyan and Kristian Skrede Gleditsch, "Refugee Flows and the Spread of Civil War," *International Organization* 60, no. 2 (2000).

60. Gleditsch, Nordås, and Salehyan, *Climate Change and Conflict*, 6.

61. Ibid.; Raleigh and Jordan, "Climate Change, Migration and Conflict."

62. Aynalem Adugna, "The 1984 Drought and Settler Migration in Ethiopia," in *Population and Disaster*, ed. John I. Clarke, et al. (Oxford: Basil Blackwell, 1989).

63. Christopher S. Clapham, *Transformation and Continuity in Revolutionary Ethiopia*, Africa Studies Series 61 (Cambridge: Cambridge University Press, 1988), Mathijs van Leeuwen, "Rwanda's Imidugudu Programme and Earlier Experiences with Villagisation and Resettlement in East Africa," *Journal of Modern African Studies* 39, no. 4 (2001).

64. Gleditsch, Nordås, and Salehyan, *Climate Change and Conflict*.

65. Intergovernmental Panel on Climate Change, *Climate Change 2007: Impacts, Adaptation and Vulnerability*; Stern, *The Economics of Climate Change*.

66. To obtain the overall index of climate change vulnerability, the study averaged each of the normalized indicators of exposure (multiple hazard risk exposure), sensitivity (human and ecological), and adaptive capacity. To identify the vulnerable areas, the study ranked the regions according to the index and divided the list into four equal parts. The researchers considered those provinces/districts falling in

the fourth quartile vulnerable areas and further classified them as mildly vulnerable, moderately vulnerable, or highly vulnerable. Arief Anshory Yusuf and Herminia Francisco, *Climate Change: Vulnerability Mapping for Southeast Asia* (Singapore: Economy and Environment Program for Southeast Asia [EEPSEA], 2009).

67. Ibid., 2.

68. Ibid., 11.

69. Ibid.

70. World Bank, *Natural Disaster Hotspots*.

71. Yusuf and Francisco, *Climate Change*, 13.

72. World Bank, *Natural Disaster Hotspots*.

73. Raleigh and Jordan, "Climate Change and Migration: Emerging Patterns in the Developing World," in *The Social Dimensions of Climate Change: Equity and Vulnerability in a Warming World*, ed. Robin Mearns and Andrew Norton (Washington, D.C.: World Bank Publications, 2009), 106.

CHAPTER 5: RIPPLE EFFECTS OF CLIMATE CHANGE MITIGATION

1. James E. Hansen, "Letter to Prime Minister Yasuo Fukuda," (Kintnersville, PA: 2008).

2. University of Portsmouth, "Iron Fertilization to Capture Carbon Dioxide Dealt a Blow: Plankton Stores Much Less Carbon Dioxide Than Estimated," *ScienceDaily*, January 29, 2009. Available at http://www.sciencedaily.com/releases/2009/01/090128183744.htm. Last accessed July 13, 2009.

3. Paul J. Crutzen, Arvin R. Mosier, Keith A. Smith, and Wilfried Winiwarter, "N_2O Release from Agro-Biofuel Production Negates Global Warming Reduction by Replacing Fossil Fuels," *Atmospheric Chemistry and Physics Discussions* 8 (2008).

4. Institute of Global Environmental Strategies, *Climate Change in the Asia-Pacific: Re-Uniting Climate Change and Sustainable Development* (Hayama, 2008), 114.

5. Tatiana Gadda and Alexandros Gasparatos, "Land Use and Cover Change in Japan and Tokyo's Appetite for Meat," *Sustainability Science*, 4, no. 2 (2009)

6. Christian Webersik and Clarice Wilson, "Achieving Environmental Sustainability and Growth in Africa: the Role of Science, Technology and Innovation," *Sustainable Development* 17, no. 6 (2009): 403.

7. FAO Corporate Document Repository, *Crop Prospects and Food Situation*, 2008.

8. Mark Thirlwell, "Food and the Spectre of Malthus," *The Financial Times*, Tuesday Feb 26, 2008. Available at http://us.ft.com/ftgateway/superpage.ft?news_id=fto022620081315470144. Last accessed July 6, 2009.

9. Joachim von Braun, *The World Food Situation New Driving Forces and Required Actions* (Washington, DC, 2007).

10. Walden Bello and Mara Baviera, "Food Wars," *Monthly Review* 61, no. 3 (2009). Available at http://www.monthlyreview.org/090706bello-baviera.php. Last accessed July 15, 2009.

11. The Earth Institute, "Global Food Crisis Golden Opportunity for African Farmers," *Press Room* (2008). Available at http://www.earth.columbia.edu/articles/view/2158. Last accessed July 10, 2009. Cited in Christian Webersik and Clarice Wilson, *Environment for African Development: A Sustainable Future through Science and Technology* (Tokyo: United Nations University, 2008), 14.

12. Miguel Esteban, Christian Webersik, David Leary, and Dexter Thompson-Pomeroy, *Innovation in Responding to Climate Change: Nanotechnology, Ocean Energy and Forestry* (Tokyo: United Nations University, 2008).

13. Intergovernmental Panel on Climate Change, *Climate Change 2007: Synthesis Report* (IPCC, 2007), 36.

14. Food and Agricultural Organisation, *State of the World's Forests 2007* (Rome: FAO, 2007).

15. Institute of Global Environmental Strategies, *Climate Change in the Asia-Pacific*.

16. Ibid., 116.

17. Ibid., 117.

18. Eviana Hartman, "A Promising Oil Alternative: Algae Energy," *The Washington Post*, January 6, 2008.

19. Ibid.

20. Frieder Hofmann, Rita Epp, Andreas Kalchschmid, Lothar Kruse, Ulrike Kuhn, Bettina Maisch, Elke Müller, et al., "GVO Pollenmonitoring zum Bt-Maisanbau im Bereich des NSG/FFH-Schutzgebietes Ruhlsdorfer Bruch," *Umweltwissenschaften Schadstoff-Forschung* 20, no. 4 (2008); Andreas Lang, Claudia Ludy, and Eva Vojtech, "Dispersion and Deposition of Bt Maize Pollen in Field Margins," *Journal of Plant Diseases and Protection* 111 (2004).

21. Alfred Wegener Institute for Polar and Marine Research, "Iron Fertilization of Oceans: A Real Option for Carbon Dioxide Reduction?" *ScienceDaily*, June 10, 2007. Available at http://www.sciencedaily.com/releases/2007/06/070608142214.htm. Last accessed July 13, 2009.

22. Sallie W. Chisholm, Paul G. Falkowski, and John J. Cullen, "Dis-Crediting Ocean Fertilization," *Science* 294, no. 5541 (2001): 309.

23. Richard Black, "Setback for Climate Technical Fix," *BBC News*, March 23, 2009.

24. Chisholm, Falkowski, and Cullen, "Dis-Crediting Ocean Fertilization."

25. John H. Martin, K.H. Coale, K.S. Johnson, S.E. Fitzwater, R.M. Gordon, S.J. Tanner, C.N. Hunter, et al., "Testing the Iron Hypothesis in Ecosystems of the Equatorial Pacific Ocean," *Nature* 371, no. 6493 (1994).

26. John Weier, "John Martin (1935–1993)," *NASA Earth Observatory, Goddard Space Flight Center* (2009). Available at http://earthobservatory.nasa.gov/Features/Martin/printall.php. Last accessed July 13, 2009.

27. Hein J.W. de Baar, Jeroen T.M. de Jong, Dorothée C.E. Bakker, Bettina M. Löscher, Cornelis Veth, Uli Bathmann, and Victor Smetacek, "Importance of Iron for Plankton Blooms and Carbon Dioxide Drawdown in the Southern Ocean," *Nature* 373 (1995).

28. Chisholm, Falkowski, and Cullen, "Dis-Crediting Ocean Fertilization."

29. Ibid., 310.

30. Ibid.

31. Ibid.

32. Michael S. Common and Sigrid Stagl, *Ecological Economics: An Introduction* (Cambridge, UK; New York: Cambridge University Press, 2005).

33. Convention on Biological Diversity, "Decisions adopted by the Conference of Parties to the Convention on Biological Diversity at its Ninth Meeting. Bonn, May 19–30, 2008" (Ninth Meeting of the Conference of the Parties to the Convention on Biological Diversity [CBD COP 9]), 96.

34. James Morgan, "Ocean Climate Fix Remains Afloat," *BBC News*, January 29, 2009. Available at http://news.bbc.co.uk/2/hi/science/nature/7856144.stm. Last accessed July 15, 2009; Alfred Wegener Institute for Polar and Marine Research, *Lohafex Provides New Insights on Plankton Ecology—Only Small Amounts of Atmospheric Carbon Dioxide Fixed* (AWI, 2009). Available at http://www.awi.de/en/news/press_releases/detail/item/lohafex_provides_new_insights_on_plankton_ecology_only_small_amounts_of_atmospheric_carbon_dioxide/?tx_list_pi1%5Bmode%5D=6&cHash=ffd0b4deee. Last accessed July 15, 2009.

35. Black, "Setback for Climate Technical Fix."

36. Ibid.

37. Alfred Wegener Institute for Polar and Marine Research, *Lohafex Provides New Insights on Plankton Ecology*.

38. Ibid.

39. Alfred Wegener Institute for Polar and Marine Research, "Iron Fertilization of Oceans"; Black, "Setback for Climate Technical Fix."

40. Alfred Wegener Institute for Polar and Marine Research, *Lohafex Provides New Insights on Plankton Ecology*.

41. William Pentland, "The Carbon Conundrum, The Environment," *Forbes Online*, October 7, 2008. Available at http://www.forbes.com/2008/10/06/carbon-sequestration-biz-energy-cx_wp_1007capture.html. Last accessed July 28, 2009.

42. Ibid.

43. Ibid.

44. Sally M. Benson and Terry Surles, "Carbon Dioxide Capture and Storage: An Overview with Emphasis on Capture and Storage in Deep Geological Formations," *Proceedings of the IEEE* 94, no. 10 (2006).

45. Ibid.

46. Ibid., 1798.

47. Ibid., 1801.

48. Ibid.

49. Ibid.

50. Bert Metz, Ogunlade Davidson, Heleen de Coninck, Manuela Loos, and Leo Meyer, *Carbon Dioxide Capture and Storage*. Cambridge: Intergovernmental Panel on Climate Change (Cambridge University Press, 2005), 25.

51. Ibid., 26.

52. Christan Klose, "Human-Triggered Earthquakes and Their Impacts on Human Security," presented to NATO *Science for Peace and Security Programme, Advanced Research Workshop "Achieving Environmental Security: Ecosystem Services and Human Welfare,"* Newport, Rhode Island Pell Center (2009).

53. Ibid.

54. Stephen Williams, "In Salah Paves the Way; A Gas Project in a Remote District of Algeria's Sahara Desert Is Pioneering Technology That Can Be Applied to Radically Cut the World's CO_2 Emissions," *The Middle East*, June 2006. Available at http://findarticles.com/p/articles/mi_m2742/is_368/ai_n24988674/. Last accessed July 28, 2009.

55. Metz, Davidson, Coninck, Loos, and Meyer, *Carbon Dioxide Capture and Storage*.

56. Benson and Surles, "Carbon Dioxide Capture and Storage," 1796.

57. Bryan Walsh, "What's Next 2008? 10 Ideas That Are Changing the World," *TIME Magazine*, March 12, 2008. Available at http://www.time.com/time/specials/2007/article/0,28804,1720049_1720050_1721653,00.html. Last accessed July 15, 2009.

58. Alan Robock, "Cooling Following Large Volcanic Eruptions Corrected for the Effect of Diffuse Radiation on Tree Rings," *Geophysical Research Letters* 32, no. 6 (2005).

59. Lianhong Gu, Dennis Baldocchi, Steve C. Wofsy, J. William Munger, Joseph J. Michalsky, Shawn P. Urbanski, and Thomas A. Boden, "Response of a Deciduous Forest to the Mount Pinatubo Eruption: Enhanced Photosynthesis," *Science* 299, no. 5615 (2003).

60. Michael C. MacCracken, *Beyond Mitigation: Potential Options for Counter-Balancing the Climatic and Environmental Consequences of the Rising Concentrations of Greenhouse Gases* (Washington, DC: The World Bank, Office of the Senior Vice President and Chief Economist, 2009), 17.

61. Ibid., 16.

62. Alan Robock, "20 Reasons Why Geoengineering May Be a Bad Idea: Carbon Dioxide Emissions Are Rising So Fast That Some Scientists Are Seriously Considering Putting Earth on Life Support as a Last Resort. But Is This Cure Worse Than the Disease?" *Bulletin of the Atomic Scientists* 64, no. 2 (2008).

63. Ibid.

64. Jamais Cascio, "Battlefield Earth," *Foreign Policy Online Edition* (2008). Available at http://www.foreignpolicy.com/story/cms.php?story_id=4146. Last accessed July 15, 2009.

65. Ibid.

66. Donald W. Olson, Russell L. Doescher, and Marilynn S. Olson, "When the Sky Ran Red: The Story Behind the Scream," *Sky & Telescope* 107, no. 2 (2004).

67. Patrick Moore, "Going Nuclear: A Green Makes the Case," *The Washington Post*, April 16, 2006.

68. Ibid.

69. Cited in Richard Black, "Nuclear University Promotes Atom Power," *BBC News Online* (2003). Available at http://news.bbc.co.uk/2/hi/science/nature/3083382.stm. Last accessed July 24, 2009.

70. Cited in Ibid.

71. Karen Bickerstaff, Irene Lorenzoni, Nick F. Pidgeon, Wouter Poortinga, and Peter Simmons, "Reframing Nuclear Power in the UK Energy Debate: Nuclear Power, Climate Change Mitigation and Radioactive Waste," *Public Understanding of Science* 17, no. 2 (2008): 146.

72. Ibid., 147.

73. Ibid.

74. Ibid., 162.

75. World Nuclear Association, *Nuclear Power Plants and Earthquakes* (London: World Nuclear Association, 2009). Available at http://www.world-nuclear.org/info/inf18.html. Last accessed July 24, 2009.

76. Martin Fackler, "Earthquake Stokes Fears over Nuclear Safety in Japan," *The New York Times, World, Asia Pacific*, July 24, 2007.

77. Ibid.

78. Dafna Linzer, "Strong Leads and Dead Ends in Nuclear Case Against Iran," *The Washington Post*, February 8, 2006.

CHAPTER 6: THE WAY FORWARD: A NEW ENVIRONMENTAL SECURITY AGENDA FOR THE TWENTY-FIRST CENTURY

1. W. Neil Adger, Shardul Agrawala, M. Monirul Qader Mirza, Cecilia Conde, Karen O'Brien, Juan Pulhin, Roger Pulwarty, et al., "Assessment of Adaptation Practices, Options, Constraints and Capacity," in *Climate Change 2007: Impacts, Adaptation and Vulnerability. Contribution of Working Group II to the Fourth Assessment Report of the Intergovernmental Panel on Climate Change*, ed. M.L. Parry, et al. (Cambridge: Cambridge University Press, 2007), 719.

2. Donald R. Nelson, W. Neil Adger, and Katrina Brown, "Adaptation to Environmental Change: Contributions of a Resilience Framework," *Annual Review of Environment and Resources* 32 (2007).

3. World Bank, *Natural Disaster Hotspots: A Global Risk Analysis* (Washington, DC: World Bank), 2005.

4. Patrick Meier, Doug Bond, and Joe Bond, "Environmental Influences on Pastoral Conflict in the Horn of Africa," *Political Geography* 26, no. 6 (2007).

5. Koko Warner, Charles Ehrhart, Alexander de Sherbinin, Susana Adamo, and Tricia Chai-Onn, *In Search of Shelter: Mapping the Effects of Climate Change on Human Migration and Displacement* (Bonn, Atlanta, New York: United Nations University Institute for Environment and Human Security, CARE International, Center for International Earth Science Information Network at the Earth Institute of Columbia University, 2009).

6. Nigel Purvis and Joshua Busby, *The Security Implications of Climate Change for the UN System* (Washington, DC, 2004).

7. Adger, Agrawala, Mirza, Conde, O'Brien, Pulhin, Pulwarty, et al., "Assessment of Adaptation Practices, Options, Constraints and Capacity."

8. Intergovernmental Panel on Climate Change, *Climate Change 2007: Synthesis Report* (IPCC, 2007), 56.

9. Richard J. T. Klein, Saleemul Huq, Fatima Denton, Thomas E. Downing, Richard G. Richels, John B. Robinson, and Ferenc L. Toth, "Inter-Relationships between Adaptation and Mitigation," in *Report of the Intergovernmental Panel on Climate Change: Climate Change 2007: Impacts, Adaptation and Vulnerability. Contribution of Working Group II to the Fourth Assessment*, ed. Martin L. Parry, et al. (Cambridge: Cambridge University Press, 2007).

10. E. Lisa F. Schipper, *Climate Change Adaptation and Development: Exploring the Linkages*, vol. 107 (Norwich, Bangkok: Tyndall Centre for Climate Change Research, School of Environmental Sciences, University of East Anglia and South East Asia START Regional Centre, 2007).

11. Intergovernmental Panel on Climate Change. *Climate Change 2007: Synthesis Report*, 65.

12. Jon Barnett, "Security and Climate Change," *Global Environmental Change* 13 (2003).

13. Ibid., 14.

14. Intergovernmental Panel on Climate Change, *Climate Change 2001: Impacts, Adaptation, and Vulnerability. Contribution of Working Group II to the Third Assessment Report of the Intergovernmental Panel on Climate Change* (Cambridge: Cambridge University Press, 2001).

15. Adger, Agrawala, Mirza, Conde, O'Brien, Pulhin, Pulwarty, et al., "Assessment of Adaptation Practices, Options, Constraints and Capacity," 720.

16. However, there are adaptation practices to respond to limited sea-level rise, for instance, efforts in the United States to acquire land that is prone to storm damages or to buffer other land. Another example is the building of higher storm surge barriers in the Netherlands, the reforestation of mangroves that act as natural barriers in the Philippines, or the conversion of farmland into salt marsh and grassland to provide natural sea defenses in the United Kingdom. See Ibid., 722.

17. Intergovernmental Panel on Climate Change, *Climate Change 2007: Synthesis Report*, 57.

18. For a full list of adaptation options/strategies and their underlying policy framework, please see ibid.

19. See also Adger, Agrawala, Mirza, Conde, O'Brien, Pulhin, Pulwarty, et al., "Assessment of Adaptation Practices, Options, Constraints and Capacity," 723.

20. Ibid.

21. Mark A. Cane, Stephen E. Zebiak, and Sean C. Dolan, "Experimental Forecasts of El Niño," *Nature* 321 (1986); Maxx Dilley, "Reducing Vulnerability to Climate Variability in Southern Africa: The Growing Role of Climate Information," *Climatic Change* 45, no. 1 (2000).

22. Adger, Agrawala, Mirza, Conde, O'Brien, Pulhin, Pulwarty, et al., "Assessment of Adaptation Practices, Options, Constraints and Capacity," 723.

23. Daniel E. Osgood, Pablo Suarez, James Hansen, Miguel Carriquiry, and Ashok Mishra, *Policy Research Working Paper 4651: Integrating Seasonal Forecasts and Insurance for Adaptation among Subsistence Farmers: The Case of Malawi* (Washington, DC: The World Bank, Development Research Group, Sustainable Rural and Urban Development Team, 2008).

24. Adger, Agrawala, Mirza, Conde, O'Brien, Pulhin, Pulwarty, et al., "Assessment of Adaptation Practices, Options, Constraints and Capacity," 721.

25. Ibid., 724.

26. Intergovernmental Panel on Climate Change. *Climate Change 2007: Synthesis Report*, 56.

27. Adger, Agrawala, Mirza, Conde, O'Brien, Pulhin, Pulwarty, et al., "Assessment of Adaptation Practices, Options, Constraints and Capacity," 734.

28. Martin Parry, Nigel Arnell, Pam Berry, David Dodman, Samuel Fankhauser, Chris Hope, Sari Kovats, et al., *Assessing the Costs of Adaptation to Climate Change: A Review of the UNFCCC and Other Recent Estimates* (London: International Institute for Environment and Development and Grantham Institute for Climate Change, 2009).

29. World Bank, *Clean Energy and Development: Towards an Investment Framework, Annex K* (Washington, DC: World Bank, 2006).

30. Nicolas Stern, *The Economics of Climate Change. The Stern Review* (London: HM Treasury, 2006), 442.

31. Parry, Arnell, Berry, Dodman, Fankhauser, Hope, Kovats, et al., *Assessing the Costs of Adaptation to Climate Change*.

32. Ibid.

33. Ibid., 14.

34. Ibid., 11.

35. Ibid.

36. Adger, Agrawala, Mirza, Conde, O'Brien, Pulhin, Pulwarty, et al., "Assessment of Adaptation Practices, Options, Constraints and Capacity," 734.

37. Stern, *The Economics of Climate Change*, 412.

38. Irene Lorenzoni, Neil F. Pidgeon, and Robert E. O'Connor, "Dangerous Climate Change: The Role for Risk Research," *Risk Analysis* 25 (2005).

39. Alan Irwin and Brian Wynne, *Misunderstanding Science: The Public Reconstruction of Science and Technology* (Cambridge: Cambridge University Press, 1996).

40. Adger, Agrawala, Mirza, Conde, O'Brien, Pulhin, Pulwarty, et al., "Assessment of Adaptation Practices, Options, Constraints and Capacity."

41. James Hansen, Sabine Marx, and Elke U. Weber, *The Role of Climate Perceptions, Expectations, and Forecasts in Farmer Decision Making: The Argentine Pampas and South Florida. IRI Technical Report 04-01* (Palisades, New York: International Research Institute for Climate Prediction, 2004). Available at http://iri.columbia.edu/outreach/publication/report/04-01/report04-01.pdf. Last accessed August 20, 2009.

42. Terre A. Satterfield, C. K. Mertz, and Paul Slovic, "Discrimination, Vulnerability, and Justice in the Face of Risk," *Risk Analysis* 24, no. 1 (2004).

43. Ibid.

44. Ibid., 115.

45. Torsten Grothmann and Anthony Patt, "Adaptive Capacity and Human Cognition: The Process of Individual Adaptation to Climate Change," *Global Environmental Change* 15 (2005).

46. Adger, Agrawala, Mirza, Conde, O'Brien, Pulhin, Pulwarty, et al., "Assessment of Adaptation Practices, Options, Constraints and Capacity."

47. Debra J. Davidson, Tim Williamson, and John R. Parkins, "Understanding Climate Change Risk and Vulnerability in Northern Forest-Based Communities," *Canadian Journal of Forest Research* 33 (2003).

48. Christian Webersik, "Fighting for the Plenty: The Banana Trade in Southern Somalia," *Oxford Development Studies* 33, no. 1 (2005).

49. Arief Anshory Yusuf and Herminia Francisco, *Climate Change: Vulnerability Mapping for Southeast Asia* (Singapore: Economy and Environment Program for Southeast Asia [EEPSEA], 2009).

50. Stern, *The Economics of Climate Change*, 433.

51. Jeff L. Brown, "High-Altitude Railway Designed to Survive Climate Change," *Civil Engineering* 75 (2005).

52. Shardul Agrawala, *Bridge over Troubled Waters: Linking Climate Change and Development* (Paris: OECD, 2005).

53. Stern, *The Economics of Climate Change*, 443.

54. Adger, Agrawala, Mirza, Conde, O'Brien, Pulhin, Pulwarty, et al., "Assessment of Adaptation Practices, Options, Constraints and Capacity," 737.

55. Frank Biermann and Ingrid Boas, "Protecting Climate Refugees: The Case for a Global Protocol," *Environment* 50, no. 6 (2008): 12.

56. United Nations Security Council, *SC/9000* (New York: United Nations, 2007).

57. Francesco Sindico, "Climate Change: A Security (Council) Issue?" *Climate Change Law Review* 1, no. 23 (2007).

58. United Nations Security Council, *SC/9000*.

59. Ibid.

60. Ibid.

61. Nicole Detraz and Michele M. Betsill, "Climate Change and Environmental Security: For Whom the Discourse Shifts," *International Studies Perspectives* 10, no. 3 (2009): 312.

62. Sindico, "Climate Change."

63. Ibid., 32.

64. Detraz and Betsill, "Climate Change and Environmental Security," 312.

65. Biermann and Boas, "Protecting Climate Refugees."

66. Ibid.

67. K. Warner, C. Ehrhart, A. de Sherbinin, S. Adamo, and T.C. Chai-Onn, *In Search of Shelter: Mapping the Effects of Climate Change on Human Migration and Displacement*. Prepared for Cooperative for Assistance and Relief Everywhere, Inc. (CARE), Atlanta, GA (2009).

68. Sindico, "Climate Change," 34.

69. United Nations Security Council, *SC/9000*.

70. Jared M. Diamond, *Collapse: How Societies Choose to Fail or Succeed* (New York: Penguin, 2006).

71. James A. Brander and Michael S. Taylor, "The Simple Economics of Easter Island: A Ricardo–Malthus Model of Renewable Resource Use," *American Economic Review* 88, no. 1 (1998); Jared M. Diamond, *Collapse: How Societies Choose to Fail or Succeed* (New York: Penguin, 2006); Rafael Reuveny and Christopher S. Decker, "Easter Island: Historical Anecdote or Warning for the Future?" *Ecological Economics* 35 (2000).

72. Mike Davis, *Late Victorian Holocausts: El Niño Famines and the Making of the Third World* (London: Verso, 2002).

73. David D. Zhang, Jane Zhang, Harry F. Lee, and Yuan-qing He, "Climate Change and War Frequency in Eastern China over the Last Millennium," *Human Ecology* 35 (2007).

74. Timothy M. Lenton, Hermann Held, Elmar Kriegler, Jim W. Hall, Wolfgang Lucht, Stefan Rahmstorf, and Hans Joachim Schellnhuber, "Tipping Elements in the Earth's Climate System," *Proceedings of the National Academy of Sciences* 105, no. 6 (2008).

75. See for more information on seasonal Arctic sea ice cover data generated by the National Snow and Ice Data Center, University of Colorado, Boulder. Available at http://nsidc.org/. Last accessed September 19, 2009.

76. Ibid.

77. United Nations, *United Nations Framework Convention on Climate Change*. New York: United Nations, 1992.

78. Jürgen Scheffran, "The Gathering Storm: Is Climate Change A Security Threat?" *Security Index* 15, no. 2 (2009): 29.

79. See UNEP/GRID-Arendal Maps and Graphics Library, Extent of Deforestation in Borneo 1950–2005, and Projection toward 2020. Available at http://maps.

grida.no/go/graphic/extent-of-deforestation-inborneo-1950-2005-and-projection-towards-2020. Last accessed June 23, 2009.

80. Intergovernmental Panel on Climate Change, *Climate Change 2007: Synthesis Report* (IPCC, 2007), 36.

81. Institute of Global Environmental Strategies, *Climate Change in the Asia-Pacific: Re-Uniting Climate Change and Sustainable Development* (Hayama, 2008).

82. ITTO, *Status of Tropical Forest Management 2005* (Yokohama: International Tropical Timber Organisation, 2006).

83. Detraz and Betsill, "Climate Change and Environmental Security," 304.

84. Ibid., 314.

85. Peter H. Liotta and Allan W. Shearer, *Gaia's Revenge: Climate Change and Humanity's Loss* (Westport, CT: Praeger Publishers, 2007), 133.

86. Detraz and Betsill, "Climate Change and Environmental Security," 307.

BIBLIOGRAPHY

"A New (under) Class of Travellers." *The Economist*, June 27, 2009.

Abdullah, Ibrahim. "Bush Path to Destruction: The Origin and Character of the Revolutionary United Front/Sierra Leone." *Journal of Modern African Studies* 36, no. 2 (1998).

Adamo, Susana B., and Alexander de Sherbinin. *The Impact of Climate Change on the Spatial Distribution of Populations and Migration, A Report Prepared for the United Nations Population Division*. Palisades: Center for International Earth Science Information Network (CIESIN), The Earth Institute at Columbia University, 2008.

Adger, W. Neil, and Timothy O'Riordan. "Population, Adaptation and Resilience." In *Environmental Science for Environmental Management*, edited by Timothy O'Riordan. Harlow: Prentice Hall, Pearson Education Limited, 1999.

Adger, W. Neil, Shardul Agrawala, M. Monirul Qader Mirza, Cecilia Conde, Karen O'Brien, Juan Pulhin, Roger Pulwarty, Barry Smit, and Kiyoshi Takahashi. "Assessment of Adaptation Practices, Options, Constraints and Capacity." In *Climate Change 2007: Impacts, Adaptation and Vulnerability. Contribution of Working Group II to the Fourth Assessment Report of the Intergovernmental Panel on Climate Change*, edited by M.L. Parry, O.F. Canziani, J.P. Palutikof, P.J. van der Linden, and C.E. Hanson, 717–43. Cambridge: Cambridge University Press, 2007.

Adugna, Aynalem. "The 1984 Drought and Settler Migration in Ethiopia." In *Population and Disaster*, edited by John I. Clarke, Peter Curson, S. L. Kayastha, and Prithvish Nag, 114–27. Oxford: Basil Blackwell, 1989.

Africa Watch and Physicians for Human Rights. "Somalia: No Mercy in Mogadishu: The Human Cost of the Conflict & the Struggle for Relief." *Africa Watch and Physicians for Human Rights* 4, no. 4 (1992).

Agrawala, Shardul. *Bridge over Troubled Waters: Linking Climate Change and Development*. Paris: OECD, 2005.

Agudelo, Paula A., and Judith A. Curry. "Analysis of Spatial Distribution in Tropospheric Temperature Trends." *Geophysical Research Letters* 31, no. L22207, DOI:10.1029/2004GL020818 (2004).

Alfred Wegener Institute for Polar and Marine Research. "Iron Fertilization Of Oceans: A Real Option For Carbon Dioxide Reduction?" *ScienceDaily*, 10 June 2007.

Alfred Wegener Institute for Polar and Marine Research. *Press Release: Lohafex Provides New Insights on Plankton Ecology—Only Small Amounts of Atmospheric Carbon Dioxide Fixed.* AWI, 2009.

Arnell, Nigel W. "Climate Change and Global Water Resources: SRES Emissions and Socio-Economic Scenarios." *Global Environmental Change* 14, no. 1 (2004).

Ashton, Peter J. "Avoiding Conflicts over Africa's Water Resources" *Ambio* 31, no. 3 (2002).

Ban Ki-moon. "A Climate Culprit in Darfur." *The Washington Post*, June 16, 2007, A15.

Barnett, Jon, and W. Neil Adger. "Climate Change, Human Security and Violent Conflict." *Political Geography* 26, no. 6 (2007).

Barnett, Jon. "Destabilizing the Environment-Conflict Thesis." *Review of International Studies* 26 (2000).

Barnett, Jon. "Security and Climate Change." *Global Environmental Change* 13 (2003).

Bates, Diane C. "Environmental Refugees? Classifying Human Migrations Caused by Environmental Change." *Population and Environment* 23, no. 5 (2002).

Battisti, David S., and Rosamond L. Naylor. "Historical Warnings of Future Food Insecurity with Unprecendented Seasonal Heat." *Science* 323 (2009).

Bello, Walden, and Mara Baviera. "Food Wars." *Monthly Review* 61, no. 3 (2009).

Benhin, James K. A. *Climate Change and South African Agriculture: Impacts and Adaptation Options*. Pretoria: The Centre for Environmental Economics and Policy in Africa, University of Pretoria, 2006.

Benson, Sally M., and Terry Surles. "Carbon Dioxide Capture and Storage: An Overview with Emphasis on Capture and Storage in Deep Geological Formations." *Proceedings of the IEEE* 94, no. 10 (2006).

Berdal, Mats R., David Malone, and International Peace Academy. *Greed & Grievance: Economic Agendas in Civil Wars*. Boulder, London, Ottawa: Lynne Rienner Publishers, International Development Research Centre, 2000.

Bickerstaff, Karen, Irene Lorenzoni, Nick F. Pidgeon, Wouter Poortinga, and Peter Simmons. "Reframing Nuclear Power in the UK Energy Debate: Nuclear Power, Climate Change Mitigation and Radioactive Waste." *Public Understanding of Science* 17, no. 2 (2008).

Biermann, Frank, and Ingrid Boas. "Protecting Climate Refugees: The Case for a Global Protocol." *Environment* 50, no. 6 (2008).

Black, Richard. *Environmental Refugees: Myth or Reality?* Geneva: UNHCR, 2001.

———. "Nuclear University Promotes Atom Power." *BBC News Online* (2003).

———. "Setback for Climate Technical Fix." *BBC News*, 23 March 2009.

Blaikie, Piers, Terry Cannon, Ian Davis, and Ben Wisner, eds. *At Risk: Natural Hazards, People's Vulnerability, and Disasters*. London: Routledge, 1994.

Boserup, Ester. *Population and Technology*. Oxford: Blackwell, 1981.

Bouwer, Laurens M., Ryan P. Crompton, Eberhard Faust, Peter Höppe, and Roger A. Pielke Jr. "Confronting Disaster Losses." *Science* 318, no. 2 (2007).

Brochmann, Marit, and Nils Petter Gleditsch. "Shared Rivers and International Cooperation." In *Paper Presented at the Workshop on Polarization and Conflict*, Nicosia (2006).

Brown, Casey, and Upmanu Lall. "Water and Economic Development: The Role of Variability and a Framework for Resilience." *Natural Resources Forum* 30, no. 4 (2006).

Brown, Jeff L. "High-Altitude Railway Designed to Survive Climate Change." *Civil Engineering* 75 (2005).

Brunborg, Helge, and Henrik Urdal. "The Demography of Conflict and Violence: An Introduction." *Journal of Peace Research* 42, no. 4 (2005).

Bryson, Bates, Zbigniew W. Kundzewicz, Shaohong Wu, and Jean Palutikof. *Climate Change and Water*. Nairobi: Intergovernmental Panel on Climate Change (IPCC), 2008.

Buhaug, Halvard, Nils Petter Gleditsch, and Ole Magnus Theisen. "Climate Change, the Environment, and Armed Conflict." Presented at the *Annual Meeting of the American Political Science Association*, Boston, MA, August 28–31, 2008.

Buhaug, Halvard, Nils Petter Gleditsch, and Ole Magnus Theisen. "Implications of Climate Change for Armed Conflict." In *The Social Dimensions of Climate Change: Equity and Vulnerability in a Warming World*, edited by Robin Mearns and Andrew Norton. Washington, DC: World Bank Publications, 2009.

Burke, Marshall B., Edward Miguel, Shanker Satyanath, John A. Dykema, and David B. Lobell, "Warming Increases the Risk of Civil War in Africa." *Proceedings of the National Academy of Sciences* 106, no. 49 (2009).

Cane, Mark A., Stephen. E. Zebiak, and Sean C. Dolan. "Experimental Forecasts of El-Niño." *Nature* 321 (1986).

Carius, Alexander, Dennis Tänzler, and Achim Maas. *Climate Change and Security: Challenges for German Development Cooperation*. Eschborn: Deutsche Gesellschaft für Technische Zusammenarbeit (GTZ) GmbH, 2008.

Cascio, Jamais. "Battlefield Earth." *Foreign Policy Online Edition* (2008).

Chan, Ngai Weng. "Choice and Constraints in Floodplain Occupation: The Influence of Structural Factors on Residential Location in Peninsular Malaysia." *Disasters* 19, no. 4 (1995).

Chatty, Dawn. *Mobile Pastoralists: Development Planning and Social Change in Oman*. New York: Columbia University Press, 1996.

Chisholm, Sallie W., Paul G. Falkowski, and John J. Cullen. "Dis-Crediting Ocean Fertilization." *Science* 294, no. 5541 (2001).

Christian Aid. *Human Tide: The Real Migration Crisis*. London: Christian Aid, 2007.

Clapham, Christopher S. *Transformation and Continuity in Revolutionary Ethiopia*, Africa Studies Series 61. Cambridge: Cambridge University Press, 1988.

Climate Change Network Nepal. "Climate Change and its Impact in Nepal." *CCNN Newsletter* Year 1, no. 1 (2007).

CNA Corporation. *National Security and the Threat of Climate Change*. Alexandria, VA: The CNA Corporation, 2008.

Coleman, James S. "Social Capital in the Creation of Human Capital." *American Journal of Sociology, Supplement Organizations and Institutions: Sociological and Economic Approaches to the Analysis of Social Structure* 94 (1988).

Collier, Paul, and Anke Hoeffler. "On the Incidence of Civil War in Africa." *Journal of Conflict Resolution* 46, no. 1 (2002).

Collier, Paul, Lani Elliott, Håvard Hegre, Anke Hoeffler, Marta Reynal-Querol, and Nicholas Sambanis. *Breaking the Conflict Trap: Civil War and Development Policy*. Washington, DC: World Bank and Oxford University Press, 2003.

Collier, Paul. *Post-Conflict Economic Recovery*. Oxford: Department of Economics, Oxford University, 2006.

Common, Michael S., and Sigrid Stagl. *Ecological Economics: An Introduction*. Cambridge, UK; New York: Cambridge University Press, 2005.

Convention on Biological Diversity. "Decisions adopted by the Conference of Parties to the Convention on Biological Diversity at its Ninth Meeting. Bonn, 19–30 May 2008." Ninth Meeting of the Conference of the Parties to the Convention on Biological Diversity [CBD COP 9].

Conze, Peter, and Thomas Labahn. "From a Socialistic System to a Mixed Economy: The Changing Framework for Somali Agriculture." In *Somalia: Agriculture in the Winds of Change*, edited by Peter Conze and Thomas Labahn, 13–54. Saarbrücken: EPI Verlag, 1986.

Crutzen, Paul J., Arvin R. Mosier, Keith A. Smith, and Wilfried Winiwarter. "N_2O Release from Agro-Biofuel Production Negates Global Warming Reduction by Replacing Fossil Fuels." *Atmospheric Chemistry and Physics* 8 (2008).

Dabelko, Geoffrey D. "Planning for Climate Change: The Security Community's Precautionary Principle." *Climatic Change* 96, no. 1–2 (2009).

Dalby, Simon. *Security and Environmental Change*. Cambridge, UK; Malden, MA: Polity, 2009.

Davidson, Debra J., Tim Williamson, and John R. Parkins. "Understanding Climate Change Risk and Vulnerability in Northern Forest-Based Communities." *Canadian Journal of Forest Research* 33 (2003).

Davis, Mike. *Late Victorian Holocausts: El Niño Famines and the Making of the Third World*. London: Verso, 2002.

de Baar, Hein J. W., Jeroen T. M. de Jong, Dorothée C. E. Bakker, Bettina M. Löscher, Cornelis Veth, Uli Bathmann, and Victor Smetacek. "Importance of Iron for Plankton Blooms and Carbon Dioxide Drawdown in the Southern Ocean." *Nature* 373 (1995).

de Sherbinin, Alexander. "A 'Human' Perspective on Four Deltas." In *GWSP-LOICZ-CSDMS Scoping Workshop on Dynamics and Vulnerability of River Delta Systems*, Boulder, Colorado, 26–28 September (2007).

———, Andrew Schiller, and Alex Pulsipher. "The Vulnerability of Global Cities to Climate Hazards." *Environment and Urbanization* 39, no. 1 (2007).

de Soysa, Indra. "Ecoviolence: Shrinking Pie or Honey Pot?" *Global Environmental Politics* 2, no. 4 (2002).

———. "Paradise Is a Bazaar? Greed, Creed, and Governance in Civil War, 1989–99." *Journal of Peace Research* 39, no. 4 (2002).

de Waal, Alexander. *Famine Crimes: Politics and the Disaster Relief Industry in Africa, African Issues*. Oxford, Bloomington: Currey, Indiana University Press, 1997.

Dearing, John. "Climate-Human-Environment Interactions: Resolving Our Past." *Climate of the Past Discussions* 2 (2006).

Detraz, Nicole, and Michele M. Betsill. "Climate Change and Environmental Security: For Whom the Discourse Shifts." *International Studies Perspectives* 10, no. 3 (2009).

Deudney, Daniel. "The Case against Linking Environmental Degradation and National Security." *Millennium* 19, no. 3 (1990).

Devereux, Stephen. *Theories of Famine.* New York, London: Harvester Wheatsheaf, 1993.

Diamond, Jared M. *Collapse: How Societies Choose to Fail or Succeed.* New York: Penguin, 2006.

Dilley, Maxx. "Reducing Vulnerability to Climate Variability in Southern Africa: The Growing Role of Climate Information." *Climatic Change* 45, no. 1 (2000).

———, Robert S. Chen, Uwe Deichmann, Arthur L. Lerner-Lam, and Margaret Arnold. *Natural Disaster Hotspots: A Global Risk Analysis, Synthesis Report.* Washington, DC: International Bank for Reconstruction and Development/ The World Bank and Columbia University, 2005.

Doos, Bo R. "Can Large-Scale Environmental Migrations Be Predicted?" *Global Environmental Change* 7 (1997).

Drysdale, John. *Stoics without Pillows: A Way Forward for the Somalilands.* London: HAAN Associates Publishing, 2000.

Ehrhart, Charles, Andrew Thow, Mark de Blois, and Alyson Warhurst. *Humanitarian Implications of Climate Change: Mapping Emerging Trends and Risk Hotspots:* UN Office for the Coordination of Humanitarian Affairs and CARE International, 2008.

Elbadawi, Ibrahim, and Nicholas Sambanis. "How Much War Will We See? Explaining the Prevalence of Civil War." *Journal of Conflict Resolution* 46, no. 3 (2002).

El-Hinnawi, Essam. *Environmental Refugees.* Nairobi: United Nations Environmental Programme, 1985.

Emanuel, Kerry A. "The Dependence of Hurricane Intensity on Climate." *Nature* 326, no. 6112 (1987).

EM-DAT. "Emergency Events Database." http://www.emdat.be.

Esteban, Miguel, Christian Webersik, David Leary, and Dexter Thompson-Pomeroy. *Innovation in Responding to Climate Change: Nanotechnology, Ocean Energy and Forestry.* Tokyo: United Nations University, 2008.

Esteban, Miguel, and Christian Webersik, "The Cost of Inaction: Estimation of the Economic Effects of an Increase in Typhoon Intensity in the Asia-Pacific Region," *International Journal of Global Warming* (2010 forthcoming).

Esteban, Miguel, Christian Webersik, and Tomoya Shibayama. "Effect of a Global Warming-Induced Increase in Typhoon Intensity on Urban Productivity in Taiwan." *Sustainability Science* 4, no. 2 (2009).

———. "Methodology for the Estimation of the Increase in Time Loss due to Future Increase in Tropical Cyclone Intensity in Japan." *Climatic Change,* DOI 10.1007/s10584-009-9725-9 (2009).

Fackler, Martin. "Earthquake Stokes Fears Over Nuclear Safety in Japan." *The New York Times, World, Asia Pacific,* July 24, 2007.

FAO Corporate Document Repository. *Crop Prospects and Food Situation,* 2008.

Faris, Stephan. "The Real Roots of Darfur." *Atlantic Monthly*, April 10, 2007.

Fearon, James D., and David D. Laitin. "Ethnicity, Insurgency and Civil War." *American Political Science Review* 97, no. 1 (2003).

Fine, Ben. "Social Capital?" *Development and Change* 30, no. 1 (1999).

Fingar, Thomas. *National Intelligence Assessment on the National Security Implications of Global Climate Change to 2030*. Washington, DC: National Intelligence Council, 2008.

Food and Agricultural Organisation. *State of the World's Forests 2007*. Rome: FAO, 2007.

Fortna, Virginia Page. "Does Peacekeeping Keep Peace? International Intervention and the Duration of Peace after Civil War." *International Studies Quarterly* 48, no. 2 (2004).

Gadda, Tatiana, and Alexandros Gasparatos. "Land Use and Cover Change in Japan and Tokyo's Appetite for Meat." *Sustainability Science* 4, no. 2 (2009).

Galtung, Johan. *The True Worlds: A Transnational Perspective*, Preferred worlds for the 1990's. New York: Free Press, 1980.

Gasana, James K. "Natural Resource Scarcity and Violence in Rwanda." In *Conserving the Peace: Resources, Livelihoods and Security*, edited by Mark Halle, Richard Matthew and Jason Switzer, 199–246. Winnipeg, Manitoba: International Institute for Sustainable Development, 2002.

German Advisory Council on Global Change. *Welt im Wandel: Sicherheitsrisiko Klimawandel*. Berlin: WBGU, 2007.

Giorgi, Filippo, Bruce Hewitson, J. Christensen, Michael Hulme, Hans Von Storch, Penny Whetton, R. Jones, Linda O. Mearns, and C. Fu. "Regional Climate Information-Evaluation and Projections." In *Climate Change 2001: The Scientific Basis*, edited by John T. Houghton, Yihui Ding, David J. Griggs, Maria Noguer, Paul J. van der Linden and Dai Xiaosu, 583–638: Cambridge University, 2001.

Gleditsch, Nils Petter, Peter Wallensteen, Mikael Eriksson, Margareta Sollenberg, and Håvard Strand. "Armed Conflict 1946–2001: A New Dataset." *Journal of Peace Research* 39, no. 5 (2002).

———, Ragnhild Nordås, and Idean Salehyan. *Climate Change and Conflict: The Migration Link: Coping with Crisis*. New York: International Peace Academy, 2007.

Gray, William M. "Global View of the Origin of Tropical Disturbances and Storms." *Monthly Weather Review* 96, no. 10 (1968).

Grothmann, Torsten, and Anthony Patt. "Adaptive Capacity and Human Cognition: The Process of Individual Adaptation to Climate Change." *Global Environmental Change* 15 (2005).

Gu, Lianhong, Dennis Baldocchi, Steve C. Wofsy, J. William Munger, Joseph J. Michalsky, Shawn P. Urbanski, and Thomas A. Boden. "Response of a Deciduous Forest to the Mount Pinatubo Eruption: Enhanced Photosynthesis." *Science* 299, no. 5615 (2003).

Hansch, Steven, Scott Lillibridge, Grace Egeland, Charles Teller, and Michael Toole. *Lives Lost, Lives Saved: Excess Mortality and the Impact of Health Interventions in the Somalia Emergency*. Washington, DC: Refugee Policy Group, 1994.

Hansen, James E. "Letter to Prime Minister Yasuo Fukuda." Kintnersville, PA, 2008.

Hansen, James, Sabine Marx, and Elke U. Weber. *The Role of Climate Perceptions, Expectations, and Forecasts in Farmer Decision Making: The Argentine Pampas and South Florida. IRI Technical Report 04–01.* Palisades, New York: International Research Institute for Climate Prediction, 2004.

Hardin, Garrett. "The Tragedy of the Commons." *Science*, no. 162 (1968).

Hartman, Eviana. "A Promising Oil Alternative: Algae Energy." *The Washington Post*, January 6, 2008.

Hauge, Wenche, and Tanja Ellingsen. "The Causal Pathway to Conflict: Beyond Environmental Scarcity." *Journal of Peace Research* 35, no. 3 (1998).

Helander, Bernhard. "The Hubeer in the Land of Plenty: Land, Labor, and Vulnerability among a Southern Somali Clan." In *The Struggle for Land in Southern Somalia: The War Behind the War*, edited by Catherine Lowe Besteman and Lee V. Cassanelli, xi, 222. London, Boulder: HAAN Associates Publishing, Westview Press, 1996.

Henderson-Sellers, Ann, Hao Zhang, Gerhard Berz, Kerry Emanuel, William Gray, Christopher W. Landsea, Greg Holland, James Lighthill, Shinn-Liang Shieh, Peter Webster, and Kendal McGuffie. "Tropical Cyclones and Global Climate Change: A Post-IPCC Assessment." *Bulletin of the American Meteorological Society* 79, no. 1 (1998).

Hendrix, Cullen S., and Sarah M. Glaser. "Trends and Triggers: Climate, Climate Change and Civil Conflict in sub-Saharan Africa." *Political Geography* 26, no. 6 (2007).

Henry, Sabine, Victor Piché, Dieudonné Ouédraogo, and Eric F. Lambin. "Descriptive Analysis of the Individual Migratory Pathways According to Environmental Typologies." *Population and Environment* 25, no. 5 (2004).

Hofmann, Frieder, Rita Epp, Andreas Kalchschmid, Lothar Kruse, Ulrike Kuhn, Bettina Maisch, Elke Müller, Steffi Ober, Jens Radtke, Ulrich Schlechtriemen, Gunther Schmidt, Winfried Schröder, Werner von der Ohe, Rudolf Vögel, Norbert Wedl, and Werner Wosniok. "GVO Pollenmonitoring zum Bt-Maisanbau im Bereich des NSG/FFH-Schutzgebietes Ruhlsdorfer Bruch." *Umweltwissenschaften Schadstoff-Forschung* 20, no. 4 (2008).

Homer-Dixon, Thomas. "The Ingenuity Gap: Can Poor Countries Adapt to Resource Scarcity?" *Population and Development Review* 21, no. 3 (1995).

———. *Environment, Scarcity, and Violence.* Princeton: Princeton University Press, 1999.

———. "Terror in the Weather Forecast." *The New York Times*, April 24, 2007.

———, and Jessica Blitt, eds. *Ecoviolence: Links among Environment, Population, and Security.* Oxford: Rowman & Littlefield, 1998.

———, and Marc A. Levy. "Correspondence: Environment and Security." *International Security* 20, no. 3 (1995).

Hugo, Graeme. "Environmental Concerns and International Migration." *International Migration Review* 30 (1996).

Humphreys, Macartan. "Natural Resources, Conflict, and Conflict Resolution: Uncovering the Mechanisms." *Journal of Conflict Resolution* 49, no. 4 (2005).

———, Jeffrey Sachs, and Joseph E. Stiglitz. *Escaping the Resource Curse.* New York: Columbia University Press, 2007.

———, and Paul Richards. *Prospects and Opportunities for Achieving the MDGs in Post-conflict Countries: A Case Study of Sierra Leone and Liberia.* New York: Center on Globalization and Sustainable Development, The Earth Institute at Columbia University, 2005.

Hunter, Lori M. "Migration and Environmental Hazards." *Population and Environment* 26, no. 4 (2005).

Huq, Saleemul, Sari Kovats, Hannah Reid, and David Satterthwaite. "Editorial: Reducing Risks to Cities from Disasters and Climate Change." *Environment and Urbanization* 19, no. 1 (2007).

Institute of Global Environmental Strategies. *Climate Change in the Asia-Pacific: Re-Uniting Climate Change and Sustainable Development.* Hayama, 2008.

Intergovernmental Panel on Climate Change. "Climate Change 2001: Impacts, Adaptation, and Vulnerability. Contribution of Working Group II to the Third Assessment Report of the Intergovernmental Panel on Climate Change." In *Climate Change 2001*, edited by James J. McCarthy, Osvaldo F. Canziani, Neil A. Leary, David J. Dokken and Kasey S. White. Cambridge: Cambridge University Press, 2001.

———. *Climate Change 2001: The Scientific Basis. Contribution of Working Group I to the Third Assessment Report of the Intergovernmental Panel on Climate Change [Core Writing Team, J.T. Houghton, Y. Ding, D.J. Griggs, M. Noguer, P.J. van der Linden, X. Dai, K. Maskell, C.A. Johnson (eds.)].* Cambridge: Cambridge University Press, 2001.

———. "Summary for Policymakers." In *Climate Change 2007: Impacts, Adaptation and Vulnerability. Contribution of Working Group II to the Fourth Assessment Report of the Intergovernmental Panel on Climate Change*, edited by Martin L. Parry, Osvaldo F. Canziani, Jean P. Palutikof, Paul J. van der Linden and Clair E. Hanson, 7–22. Cambridge, UK: Cambridge University Press, 2007.

———. *Climate Change 2007: Synthesis Report. Contribution of Working Groups I, II and III to the Fourth Assessment Report of the Intergovernmental Panel on Climate Change [Core Writing Team, Pachauri, Rajendra K. and Reisinger, Andy (eds.)].* Geneva, Switzerland: IPCC, 2007.

International Centre for Integrated Mountain Development. "Flash Floods in the Himalayas." *ICIMOD* (2008).

Irwin, Alan, and Brian Wynne. *Misunderstanding Science: The Public Reconstruction of Science and Technology.* Cambridge: Cambridge University Press, 1996.

ITTO. *Status of Tropical Forest Management 2005.* Yokohama: International Tropical Timber Organisation, 2006.

Jackson, Sukhan, and Adrian C. Sleigh. "Resettlement for China's Three Gorges Dam: Socio-Economic Impact and Institutional Tensions." *Communist and Post-Communist Studies* 33 (2000).

Kabir, M. M., Bibhas Chandra Saha, and Jalauddin M. Abdul Hye. *Cyclonic Storm Surge Modelling for Design of Coastal Polder.* Dhaka, Bangladesh: Institute of Water Modelling, 2009.

Kandji, Serigne, Louis Verchot, and Jens Mackensen. *Climate Change and Variability in the Sahel Region: Impacts and Adaptation Strategies in the Agricultural Sector.* Nairobi: United Nations Environmental Programme and World Agroforestry Center, 2006.

Kaplan, Robert D. "The Coming Anarchy." *Atlantic Monthly* 2 (1994).

Kasperson, Jeanne X., Roger E. Kasperson, and Billie L. Turner II, eds. *Regions at Risk: Comparisons of Threatened Environments.* Tokyo: United Nations University Press, 1995.

Kassa, Alemayehu. "Drought Risk Monitoring for the Sudan." SOAS, 1999.

Keen, David. "Incentives and Disincentives for Violence." In *Greed & Grievance: Economic Agendas in Civil Wars*, edited by Mats R. Berdal, David Malone and International Peace Academy, vii, 251. Boulder, London, Ottawa: Lynne Rienner Publishers, International Development Research Centre, 2000.

Kelly, Mick, and Neil Adger. *Assessing Vulnerability to Climate Change and Facilitating Adaptation.* Norwich: Centre for Social and Economic Research on the Global Environment, University of East Anglia, Norwich, and University College London, 1999.

Klein, Richard J. T., Saleemul Huq, Fatima Denton, Thomas E. Downing, Richard G. Richels, John B. Robinson, and Ferenc L. Toth. "Inter-Relationships between Adaptation and Mitigation." In *Report of the Intergovernmental Panel on Climate Change: Climate Change 2007: Impacts, Adaptation and Vulnerability. Contribution of Working Group II to the Fourth Assessment,* edited by Martin L. Parry, Osvaldo F. Canziani, Jean P. Palutikof, Paul J. van der Linden, and Clair E. Hanson, 745–77. Cambridge: Cambridge University Press, 2007.

Klose, Christian. "Human-Triggered Earthquakes and Their Impacts on Human Security." Presented to *NATO Science for Peace and Security Programme, Advanced Research Workshop "Achieving Environmental Security: Ecosystem Services and Human Welfare,"* Newport, Rhode Island, Pell Center (2009).

———, and Christian Webersik. "Assessing the Inter-Relationship Between Geo-hazards, Sustainable Development, and Political Stability: A Comparative Study for Haiti and the Dominican Republic." Presented at the *7th International Science Conference on the Human Dimensions of Global Environmental Change (Open Meeting 2009).* Bonn, 2009.

Knutson, Thomas R., Robert E. Tuleya, Weixing Shen, and Isaac Ginis. "Impact of CO_2-Induced Warming on Hurricane Intensities as Simulated in a Hurricane Model with Ocean Coupling." *Journal of Climate* 14 (2001).

Lacina, Bethan, and Nils Petter Gleditsch. "Monitoring Trends in Global Combat: A New Dataset of Battle Deaths." *European Journal of Population* 21, no. 2–3 (2005).

Landsea, Christopher W., Bruce A. Harper, Karl Hoarau, and John A. Knaff. "Can We Detect Trends in Extreme Tropical Cyclones?" *Science* 313, no. 5786 (2006).

Lang, Andreas, Claudia Ludy, and Eva Vojtech. "Dispersion and Deposition of Bt Maize Pollen in Field Margins." *Journal of Plant Diseases and Protection* 111 (2004).

Larsen, Janet. *Setting the Record Straight: More Than 52,000 Europeans Died from Heat in Summer 2003*: Earth Policy Institute, 2006.

Lenton, Timothy M., Hermann Held, Elmar Kriegler, Jim W. Hall, Wolfgang Lucht, Stefan Rahmstorf, and Hans Joachim Schellnhuber. "Tipping Elements in the Earth's Climate System." *Proceedings of the National Academy of Sciences* 105, no. 6 (2008).

Levy, Marc A. "Is the Environment a National Security Issue?" *International Security* 20, no. 2 (1995).

———, Catherine Thorkelson, Charles Vörösmarty, Ellen Douglas, and Macartan Humphreys. "Freshwater Availability Anomalies and Outbreak of Internal War: Results from a Global Spatial Time Series Analysis." In *Human Security and Climate Change,* An International Workshop Holmen Fjord Hotel, Asker, near Oslo, 21–23 June (2005).

Lind, Jeremy, and Kathryn Sturman, eds. *Scarcity and Surfeit: The Ecology of Africa's Conflicts.* Pretoria: Institute for Security Studies, 2002.

Linzer, Dafna. "Strong Leads and Dead Ends in Nuclear Case against Iran." *The Washington Post*, February 8, 2006.

Liotta, Peter H., and Allan W. Shearer. *Gaia's Revenge: Climate Change and Humanity's Loss*. Westport, Conn.: Praeger Publishers, 2007.

Lipschutz, Ronnie D. "Environmental Conflict: A Values-Oriented Approach." In *Conflict and the Environment*, edited by Nils Petter Gleditsch. Dordrecht: Kluwer Academic, 1997.

Little, Peter D. *Somalia: Economy without State*. Oxford, Bloomington & Indianapolis, Hargeisa: James Currey, Indiana University Press, Btec Books, 2003.

Lonergan, Stephen C. *The Role of Environmental Degradation in Population Displacement*. Washington, DC: Environmental Change and Security Program, Woodrow Wilson Center, 1998.

Lorenzoni, Irene, Neil F. Pidgeon, and Robert E. O'Connor. "Dangerous Climate Change: The Role for Risk Research." *Risk Analysis* 25 (2005).

MacCracken, Michael C. *Beyond Mitigation: Potential Options for Counter-Balancing the Climatic and Environmental Consequences of the Rising Concentrations of Greenhouse Gases*. Washington, DC: The World Bank, Office of the Senior Vice President and Chief Economist, 2009.

McGranahan, Gordon, Deborah Balk, and Bridget Anderson. "The Rising Tide: Assessing the Risks of Climate Change and Human Settlements in Low Elevation Coastal Zones." *Environment and Urbanization* 19, no. 1 (2007).

McLeman, Robert, and Barry Smit. "Migration as an Adaptation to Climate Change." *Climatic Change* 76 (2006).

Malthus, T. Robert. *First Essay on the Principle of Population*. London: Oxford University Press, 1798.

———, and James Bonar. *First Essay on Population, 1798*. London, New York: Macmillan, St. Martin's Press, 1966.

Martin, John H., K.H. Coale, K.S. Johnson, S.E. Fitzwater, R.M. Gordon, S.J. Tanner, C.N. Hunter, V.A. Elrod, J.L. Nowicki, T.L. Coley, R.T. Barber, S. Lindley, A.J. Watson, K. Van Scoy, C.S. Law, M.I. Liddicoat, R. Ling, T. Stanton, J. Stockel, C. Collins, A. Anderson, R. Bidigare, M. Ondrusek, M. Latasa, F.J. Millero, K. Lee, W. Yao, J.Z. Zhang, G. Friederich, C. Sakamoto, F. Chavez, K. Buck, Z. Kolber, R. Greene, P. Falkowski, S.W. Chisholm, F. Hoge, R. Swift, J. Yungel, S. Turner, P. Nightingale, A. Hatton, P. Liss, and N.W. Tindale. "Testing the Iron Hypothesis in Ecosystems of the Equatorial Pacific Ocean." *Nature* 371, no. 6493 (1994).

Matthew, Richard A. "Environment, Population and Conflict." *Journal of International Affairs* 56, no. 1 (2002).

Meier, Patrick, Doug Bond, and Joe Bond. "Environmental Influences on Pastoral Conflict in the Horn of Africa." *Political Geography* 26, no. 6 (2007).

Metz, Bert, Ogunlade Davidson, Heleen de Coninck, Manuela Loos, and Leo Meyer. *Carbon Dioxide Capture and Storage*. Cambridge: Intergovernmental Panel on Climate Change; Cambridge University Press, 2005.

Miguel, Edward, Shanker Satyanath, and Ernest Sergenti. "Economic Shocks and Civil Conflict: An Instrumental Variables Approach." *Journal of Political Economy* 112, no. 4 (2004).

Miller, Gifford H., Marilyn L. Fogel, John W. Magee, Michael K. Gagan, Simon J. Clarke, and Beverly J. Johnson. "Ecosystem Collapse in Pleistocene Australia and a Human Role in Megafaunal Extinction." *Science* 309 (2005).

Moore, Patrick. "Going Nuclear: A Green Makes the Case." *The Washington Post,* April 16, 2006.

Morgan, James. "Ocean Climate Fix Remains Afloat." *BBC News,* January 29, 2009.

Morris, Saul S., Oscar Neidecker-Gonzales, Calogero Carletto, Marcial Munguía, Juan Manuel Medina, and Quentin Wodon. "Hurricane Mitch and the Livelihoods of the Rural Poor in Honduras." *World Development* 30, no. 1 (2002).

Mountain Forum. "Natural Resources: Women, Conflicts and Management." *Mountain Forum Bulletin* 8, no. 2 (2008).

Mukhtar, Mohamed Haji. "The Plight of the Agro-Pastoral Society of Somalia." *Review of African Political Economy* 23, no. 70 (1996).

Munich Re Group. *Katrina and Rita: Munich Re Estimates Total Insured Market Losses at Up to US$ 40bn.* Munich: Munich Re Group, 2005.

———. *Topics Geo: Natural Catastrophes 2006.* Munich: Munich Re Group, 2007.

Murshed, S. Mansoob, and Scott Gates. "Spatial-Horizontal Inequality and the Maoist Insurgency in Nepal." *Review of Development Economics* 9, no. 1 (2005).

Mutz, Reinhard, and Bruno Schoch. *Friedensgutachten.* Münster: LIT Verlag, 1995.

Myers, Norman. "Environmental Refugees." *Population and Environment: A Journal of Interdisciplinary Studies* 19 (1997).

———. *Environmental Exodus: An Emergent Crisis in the Global Arena.* Washington, DC: Climate Institute, 1995

National Intelligence Council. *Global Trends 2025: A Transformed World; NIC 2008–003.* Washington, DC, 2008

Nelson, Donald R., W. Neil Adger, and Katrina Brown. "Adaptation to Environmental Change: Contributions of a Resilience Framework." *Annual Review of Environment and Resources* 32 (2007).

Nils Petter Gleditsch, Kathryn Furlong, Håvard Hegre, Bethany Lacina, and Taylor Owen. "Conflicts over Shared Rivers: Resource Scarcity or Fuzzy Boundaries?" *Political Geography* 25, no. 4 (2006).

Nordås, Ragnhild, and Nils Petter Gleditsch. "Climate Change and Conflict." *Political Geography* 26, no. 6 (2007).

Nordhaus, William D. *The Economics of Hurricanes in the United States.* Boston: Annual Meetings of the American Economic Association, 2006.

Olson, Donald W., Russell L. Doescher, and Marilynn S. Olson. "When the Sky Ran Red: The Story Behind the Scream." *Sky & Telescope* 107, no. 2 (2004).

Opletal, Helmut. "'Nicht der Krieg schuf das Desaster in Somalia' Ein Bericht von Africa Watch nennt die Ursachen für Hunger und Krieg am Horn von Afrika." *Frankfurter Rundschau,* May 14, 1993, 12.

Osgood, Daniel E., Pablo Suarez, James Hansen, Miguel Carriquiry, and Ashok Mishra. *Policy Research Working Paper 4651: Integrating Seasonal Forecasts and Insurance for Adaptation among Subsistence Farmers: The Case of Malawi.* Washington, DC: The World Bank, Development Research Group, Sustainable Rural and Urban Development Team, 2008

Paris, Roland. "Human Security: Paradigm Shift or Hot Air?" *International Security* 26, no. 2 (2001).

Parry, Martin, Nigel Arnell, Pam Berry, David Dodman, Samuel Fankhauser, Chris Hope, Sari Kovats, Robert Nicholls, David Satterthwaite, Richard Tiffin, and

Tim Wheeler. *Assessing the Costs of Adaptation to Climate Change: A Review of the UNFCCC and Other Recent Estimates*. London: International Institute for Environment and Development and Grantham Institute for Climate Change, 2009.

Parry, Martin, Osvaldo Canziani, Jean Palutikof, Paul van der Linden, and Clair Hanson. *Climate Change 2007: Impacts, Adaptation and Vulnerability: Summary for Policymakers and Technical Summary*: WMO, UNEP, 2007.

Pearce, David William, and R. Kerry Turner. *Economics of Natural Resources and the Environment*. Hemel Hempstead: Harvester Wheatsheaf, 1990.

Pentland, William. "The Carbon Conundrum, The Environment." *Forbes Online*, 7 October 2008.

Perch-Nielsen, Sabine. "Understanding the Effect of Climate Change on Human Migration: The Contribution of Mathematical and Conceptual Models." Swiss Federal Institute of Technology, 2004.

Pirard, Philippe, Stéphanie Vandentorren, Mathilde Pascal, Karine Laaidi, Alain Le Tertre, Sylvie Cassadou, and Martine Ledrans. "Summary of the Mortality Impact Assessment of the 2003 Heat Wave in France." *Eurosurveillance* 10, no. 7 (2005).

Population Division of the Department of Economic and Social Affairs of the United Nations Secretariat. *World Population Prospects: The 2005 Revision*. New York, 2005.

———. *World Population Prospects: The 2006 Revision*. New York, 2006.

Purvis, Nigel, and Joshua Busby. *The Security Implications of Climate Change for the UN System*. Washington, DC, 2004.

Raleigh, Clionadh, and Henrik Urdal. "Climate Change, Environmental Degradation and Armed Conflict." *Political Geography* 26, no. 6 (2007).

———, and Lisa Jordan. "Climate Change, Migration and Conflict." Presented at the *International Studies Association annual convention*. San Francisco, CA, 26–29 March, 2007.

———. and Lisa Jordan. "Climate Change and Migration: Emerging Patterns in the Developing World." In *The Social Dimensions of Climate Change: Equity and Vulnerability in a Warming World*, edited by Robin Mearns and Andrew Norton. Washington, DC: World Bank Publications, 2009.

Refugee Policy Group. *Hope Restored? Humanitarian Aid in Somalia, 1990–1994*. Washington, DC: Refugee Policy Group, 1994.

Richards, Paul. *Fighting for the Rain Forest: War, Youth & Resources in Sierra Leone, African Issues*. Oxford, Portsmouth, NH: James Currey, Heinemann, 1996.

Robock, Alan. "20 Reasons Why Geoengineering May Be a Bad Idea: Carbon Dioxide Emissions Are Rising So Fast That Some Scientists Are Seriously Considering Putting Earth on Life Support as a Last Resort. But Is This Cure Worse Than the Disease?" *Bulletin of the Atomic Scientists* 64, no. 2 (2008).

Robock, Alan. "Cooling Following Large Volcanic Eruptions Corrected for the Effect of Diffuse Radiation on Tree Rings." *Geophysical Research Letters* 32, no. 6 (2005).

Ross, Eric B. *The Malthus Factor: Population, Poverty and Politics in Capitalist Development*. London: Zed, 1998.

Sachs, Jeffrey D. "The Strategic Significance of Global Inequality." *The Washington Quarterly* 24, no. 3 (2001).

————. *Common Wealth: Economics for a Crowded Planet.* New York: Penguin Press, 2008.

Salehyan, Idean, and Kristian Skrede Gleditsch. "Refugee Flows and the Spread of Civil War." *International Organization* 60, no. 2 (2000).

Satterfield, Terre A., C. K. Mertz, and Paul Slovic. "Discrimination, Vulnerability, and Justice in the Face of Risk." *Risk Analysis* 24, no. 1 (2004).

Scheffran, Jürgen. "Climate Change and Security: How Is Global Warming Affecting Existing Competition for Resources and Changing International Security Priorities? A Survey of Recent Research Shows How Complex the Picture Could Become." *Bulletin of the Atomic Scientists* 64, no. 2 (2008).

————. "The Gathering Storm: Is Climate Change A Security Threat?" *Security Index* 15, no. 2 (2009).

Schipper, E. Lisa F. *Climate Change Adaptation and Development: Exploring the Linkages.* Vol. 107. Norwich, Bangkok: Tyndall Centre for Climate Change Research, School of Environmental Sciences, University of East Anglia and South East Asia START Regional Centre, 2007.

Schwartz, Peter, and Doug Randall. *An Abrupt Climate Change Scenario and Its Implications for United States National Security.* New York: Environmental Defense Fund, 2003.

Sen, Amartya Kumar. "Food, Economics and Entitlements." In *The Political Economy of Hunger: Selected Essays*, edited by Jean Drèze, Amartya Kumar Sen, and Athar Hussain. Oxford: Clarendon Press, 1995.

————. "Ingredients of Famine Analysis: Availability and Entitlements." *Quarterly Journal of Economics* 96, no. 3 (1981).

Shapiro, Lloyd J., and Stanley B. Goldenberg. "Atlantic Sea Surface Temperatures and Tropical Cyclone Formation." *Journal of Climate* 11, no. 4 (1998).

Shrestha, Arun B., Cameron P. Wake, Paul A. Mayewski, and Jack E. Dibb. "Maximum Temperature Trends in the Himalaya and Its Vicinity: An Analysis Based on Temperature Records from Nepal for the Period 1971–94." *Journal of Climate* 12 (1999).

Sindico, Francesco. "Climate Change: A Security (Council) Issue?" *Climate Change Law Review* 1, no. 23 (2007).

State Failure Task Force. *State Failure Task Force Report: Phase II Findings* (No. 5). Washington, DC: Woodrow Wilson Center, 1999.

Stedman, Stephen. "Spoiler Problems in Peace Processes." *International Security* 22, no. 2 (1997).

Stern, Nicolas. *The Economics of Climate Change. The Stern Review.* London: HM Treasury, 2006.

Stromberg, Per, Miguel Esteban, and Dexter Thompson-Pomeroy. *Interlinkages in Climate Change: Vulnerability of a Mitigation Strategy? Impact of Increased Typhoon Intensity on Biofuel Production in the Philippines.* Tokyo, 2009.

Tata Energy Research Institute. *The Economic Impact of a One-Metre Sea-Level Rise on the Indian Coastline: Method and Case Studies, Report Submitted to the Ford Foundation.* Delhi: TERI, 1996.

Thapa, Manish. "Maoist Insurgency in Nepal: Context, Cost and Consequences." In *Afro-Asian Conflicts*, edited by Seema Shekhawat and Debidatta Aurobinda Mahapatra. New Delhi: New Century Publications, 2008.

The Earth Institute. "Global Food Crisis Golden Opportunity for African Farmers." *Press Room* (2008).

Theisen, Ole Magnus, Helge Holtermann, and Halvard Buhaug. "Drought, Political Exclusion, and Civil War." Presented at the *International Studies Association annual convention*. New Orleans, LA, 17–20 February, 2010.

Thirlwell, Mark. "Food and the Spectre of Malthus." *The Financial Times*, February
26, 2008.

Tiffen, Mary, Michael Mortimore, and Francis Gichuki. *More People, Less Erosion:
Environmental Recovery in Kenya*. Chichester: Wiley, 1994.

United Nations. *United Nations Framework Convention on Climate Change*. New
York: United Nations, 1992.

United Nations Development Programme. *Human Development Report, 1994*. New
York: Oxford University Press, 1994.

———. *Human Development Report, Somalia 2001*. Nairobi: United Nations Development Programme Somalia Country Office, 2001.

———. *Human Development Report 2007/2008, Fighting Climate Change: Human
Solidarity in a Divided World*. New York: UNDP. (Palgrave Macmillan), 2007.

United Nations Economic Commission for Africa. *Economic Report on Africa*.
Addis Ababa: UNECA, 2005.

United Nations Environment Programme. *Marine and Coastal Ecosystems and
Human Well-Being: A Synthesis Report Based on the Findings of the Millennium
Ecosystem Assessment*. Nairobi: United Nations Environment Programme,
2006.

———. *Fourth Global Environment Outlook: Environment for Development Assessment Report*. Nairobi: United Nations Environment Programme, 2007.

———. *Sudan Post-Conflict Environmental Assessment*. Nairobi: United Nations
Environment Programme, 2007.

UNFCCC. *Report of the Conference of the Parties on Its Eleventh Session, Held at
Montreal from November 28 to December 10, 2005. Part One: Proceedings*. In
UN Doc. FCCC/CP/2005/5. Bonn: United Nations Framework Convention
on Climate Change, 2006.

UNFCCC. *Report of the Conference of the Parties on Its Thirteenth Session, Held in
Bali from December 3 to 15, 2007*. Bonn: United Nations Convention on Climate Change, 2008.

United Nations Population Division. *World Population Prospects: The 2002 Revision*.
New York: United Nations Population Division, 2003.

United Nations Security Council. *SC/9000*. New York: United Nations, 2007.

University of Portsmouth. "Iron Fertilization To Capture Carbon Dioxide Dealt a
Blow: Plankton Stores Much Less Carbon Dioxide than Estimated." *ScienceDaily*, January 29, 2009.

Urdal, Henrik. "People vs. Malthus: Population Pressure, Environmental Degradation and Armed Conflict Revisited." *Journal of Peace Research* 42, no. 4
(2005).

van Leeuwen, Mathijs. "Rwanda's Imidugudu Programme and Earlier Experiences
with Villagisation and Resettlement in East Africa." *Journal of Modern
African Studies* 39, no. 4 (2001).

von Braun, Joachim. *The World Food Situation New Driving Forces and Required Actions*. Washington, DC, 2007.

————, and Ruth Meinzen-Dick. *"Land-Grabbing" by Foreign Investors in Developing Countries: Risks and Opportunities*. Washington, DC, 2009.

Walsh, Bryan. "What's Next 2008? 10 Ideas That Are Changing the World." *TIME Magazine*, March 12, 2008.

Walter, Barbara F. "Does Conflict Beget Conflict? Explaining Recurrence in Civil War." *Journal of Peace Research* 41, no. 3 (2004).

Ware, Helen. "Demography, Migration and Conflict in the Pacific." *Journal of Peace Research* 42, no. 4 (2005).

Warner, Koko, Charles Ehrhart, Alexander de Sherbinin, Susana Adamo, and Tricia Chai-Onn. *In Search of Shelter: Mapping the Effects of Climate Change on Human Migration and Displacement*. Bonn, Atlanta, New York: United Nations University Institute for Environment and Human Security, CARE International, Center for International Earth Science Information Network at the Earth Institute of Columbia University, 2009.

Warren, Rachel, Nigel Arnell, Robert Nicholls, Peter Levy, and Jeff Price. *Understanding the Regional Impacts of Climate Change*: Research Report Prepared for the *Stern Review on the Economics of Climate Change*. Norwich: Tyndall Centre, 2006.

Webersik, Christian. "Differences that Matter: The Struggle of the Marginalised in Somalia." *Africa* 74, no. 4 (2004).

————. "Fighting for the Plenty: The Banana Trade in Southern Somalia." *Oxford Development Studies* 33, no. 1 (2005).

————. "Wars over Resources? Evidence from Somalia." *Environment* 50, no. 3 (2008).

————, and Clarice Wilson. *Environment for African Development: A Sustainable Future through Science and Technology*. Tokyo: United Nations University, 2008.

————. "Achieving Environmental Sustainability and Growth in Africa: the Role of Science, Technology, and Innovation." *Sustainable Development* 17, no. 6 (2009).

————, Miguel Esteban, and Tomoya Shibayama. "The Economic Impact of Future Increase in Tropical Cyclones in Japan." *Natural Hazards*, DOI: 10.1007/s11069-010-9522-9 (2010).

Webster, Peter, Greg Holland, Judith A. Curry, and H.-R. Chang. "Changes in Tropical Cyclone Number, Duration, and Intensity in a Warming Environment." *Science* 309, no. 5742 (2005).

Weier, John. "John Martin (1935–1993)." *NASA Earth Observatory, Goddard Space Flight Center* (2009).

Williams, Stephen. "In Salah Paves the Way; A Gas Project in a Remote District of Algeria's Sahara Desert is Pioneering Technology That Can Be Applied to Radically Cut the World's CO_2 Emissions." *The Middle East*, June 2006.

Wisner, Ben. "Jilaal, Gu, Hagaa, and Der: Living with the Somali Land, and Living Well." In *The Somali Challenge: From Catastrophe to Renewal?* Edited by Ahmed I. Samatar, 29–59. Boulder, London: Lynne Rienner, 1994.

Wolf, Aaron T. "Conflict and Cooperation along International Waterways." *Water Policy* 1, no. 1 (1998).

———. "Water Wars and Water Reality: Conflict and Cooperation along International Waterways." In *Environmental Change, Adaptation and Human Security*, edited by Steve Lonergan, 251–65. Dordrecht: Kluwer, 1999.

Woodham-Smith, Cecil. *The Great Hunger: Ireland 1845–1849*. London: Penguin, 1991.

World Bank. *Natural Disaster Hotspots: A Global Risk Analysis*. Washington, DC: World Bank, 2005.

———. *Clean Energy and Development: Towards an Investment Framework, Annex K*. Washington, DC: World Bank, 2006.

———. *World Development Indicators*. Washington, DC: The World Bank, 2007.

World Nuclear Association. "Nuclear Power Plants and Earthquakes." London: World Nuclear Association, 2009.

World Wide Fund for Nature. *Glaciers, Glacier Retreat, and its Subsequent Impacts in Nepal, India and China*. Kathmandu: WWF Nepal Program, 2005.

Yusuf, Arief Anshory, and Herminia Francisco. *Climate Change: Vulnerability Mapping for Southeast Asia*. Singapore: Economy and Environment Program for Southeast Asia (EEPSEA), 2009.

Zaman, M.Q. "The Displaced Poor and Resettlement Policies in Bangladesh." *Disasters* 15, no. 2 (1991).

Zellen, Barry Scott. Arctic Doom, Arctic Boom the Geopolitics of Climate Change in the Arctic. Santa Barbara, CA: Praeger, 2009.

Zhang, David D., Jane Zhang, Harry F. Lee, and Yuan-qing He. "Climate Change and War Frequency in Eastern China over the Last Millennium." *Human Ecology* 35 (2007).

INDEX

About the Author

CHRISTIAN WEBERSIK is currently working as Associate Professor at the Centre for Development Studies at the University of Agder (UiA) in Norway. His general research interests are climate change and security, natural hazards and development, and post-conflict economic recovery. Before joining UiA, he was a Japan Society for the Promotion of Science—United Nations University (JSPS-UNU) Postdoctoral Fellow at United Nations University's Institute of Advanced Studies (UNU-IAS). Webersik briefly worked as report writer for United Nations Development Programme (UNDP) Bureau for Crisis Prevention and Recovery. Before that, he was a fellow at Columbia University's Earth Institute where he was hosted by the Center for International Earth Science Information Network (CIESIN). Following his doctorate, he was Assistant Professor of political science at Asmara University, Eritrea. He holds a D.Phil. in Politics and International Relations from Oxford University, where he studied the political economy of war and the role of natural resources in conflict in Somalia. In the past, Webersik worked in a number of conflict situations with UNDP and the United Nations Office for the Coordination of Humanitarian Affairs. He also worked for the UN Climate Change Secretariat (UNFCCC) in Bonn, Germany. He continues to be interested in understanding in how humans interact with their environment, to what extent environmental factors play a role in armed conflict, and the impact of natural hazards on people's well-being and livelihoods.